SPACE SCIENCE, EXPLORATION AND POLICIES

METEORITES AND ASTEROIDS

CLASSIFICATION, GEOLOGY AND EXPLORATION

SPACE SCIENCE, EXPLORATION AND POLICIES

Additional books in this series can be found on Nova's website under the Series tab.

Additional e-books in this series can be found on Nova's website under the e-books tab.

SPACE SCIENCE, EXPLORATION AND POLICIES

METEORITES AND ASTEROIDS

CLASSIFICATION, GEOLOGY AND EXPLORATION

AKILINA DEMENTIEVA
AND
DANILO OSTROGORSKY
EDITORS

Nova Science Publishers, Inc.
New York

For permission to use material from this book please contact us:
Telephone 631-231-7269; Fax 631-231-8175
Web Site: http://www.novapublishers.com

NOTICE TO THE READER

LIBRARY OF CONGRESS CATALOGING-IN-PUBLICATION DATA

Meteorites and asteroids : classification, geology, and exploration / editors, Akilina Dementieva and Danilo Ostrogorsky.
p. cm.
Includes bibliographical references and index.
ISBN 978-1-61942-115-8 (hardcover)
1. Meteorites. 2. Asteroids. I. Dementieva, Akilina. II. Ostrogorsky, Danilo.
QB755.M484 2011
523.5'1--dc23
2011043197

Published by Nova Science Publishers, Inc. † New York

CONTENTS

PREFACE

In this book, the authors present current research in the study of the classification, geology and exploration of asteroids and meteorites. Topics discussed include meteorites and their asteroidal parent bodies; the diversion and exploitation of ice-rich NEOs using the solar collector; radar characteristics of asteroid 33342 (1998 WT24); asteroid dimensions and the truncated pareto distribution; Hilda asteroids in the Jupiter neighborhood; and asteroid Apophis and 1950 DA.

Chapter 1 - Meteorites are fragments of planetary objects, most often of asteroids, that have broken off of their parent bodies and made their way to Earth. For years, one of the primary goals of planetary scientists who study these objects has been to determine possible connections between recovered meteorites and their potential parent bodies in the asteroid belt and near-Earth asteroid population. Mineralogies and chemistries of meteorites are determined in laboratories, using traditional analytical techniques such as X-ray diffraction and electron microprobe phase analysis. This, along with oxygen isotopic analysis, has resulted in the identification of approximately twenty-five different groups of stony and stony-iron meteorites (those that contain silicate mineral and Fe-Ni metal) and thirteen groups of iron meteorites. Iron and stony-iron meteorites are derived from differentiated asteroids, or those that have a core, mantle, and a crust. Each meteorite group is usually thought to be derived from a different parent body.

The majority of stony meteorites originate from primitive asteroids that did not experience melting, though a smaller number are pieces of crust from differentiated parent bodies. Because of their remote location, asteroids cannot be studied using the analytical techniques applied to meteorites. Instead, they are studied using high-powered telescopes that record the amount of light that is reflected from the asteroid surface. Many minerals preferentially absorb light at particular wavelengths, resulting in characteristic absorption features for these minerals. Visible/near-infrared spectra have been the most widely-applied tool used in the search for these parent asteroids, due to the strong 1 μm and/or 2 μm absorption bands present in the abundant minerals olivine and pyroxene. Abundances and compositions of olivine and pyroxene can be derived by characterizing these spectral features. Comparison of meteorite and asteroid mineralogy has led to linkages between several meteorite groups and possible parent bodies, including two very likely meteorite-asteroid relationships between S-type asteroids and the ordinary chondrites and asteroid 4 Vesta and the HED (howardite, eucrite, and diogenite) meteorites. Though uncertainties in optical, chemical, and physical properties of asteroid surfaces make it difficult to decisively link most asteroids with particular meteorite groups, the search for meteorite parent bodies continues to dominate asteroid spectroscopic studies.

Chapter 2 - The process of extraterrestrial material income into the Earth's atmosphere is considered. It is shown that particles of meteoric smoke may serve as condensation nuclei in the troposphere and stratosphere. It is shown that charged particles of meteoric origin might appreciably change mesospheric, and, as a result, the total atmospheric resistance, which, in turn, affect the global current circuit. Changes in the global electric circuit may influence processes of aerosol and cloud formation. The possible climatic effect of interstellar dust in the Earth's atmosphere is found negligible. Possible connection between terrestrial climatic variation with a period 2.0-2.5 kyr (the Hallstatt cycle) and nodal precession of the core of Taurid meteoroid stream is analyzed. It is confirmed that extremes of the Hallstatt cycle coincide with the data of probable intersections of the Earth's orbit by the core of Taurid complex. Since during these epochs intensity of the Earth's bombardment by cometary debris increases sharply, the possible results of collision of the Earth with asteroid, which has size of 250 m and velocity of 20 $km \times s^{-1}$ is estimated. It is shown that climatic consequences of such impact might be substantial and therefore a chain of such events may contribute to 2.0-2.5 kyr climatic variation appreciably. The obtained results testify that the extraterrestrial factors actually might contribute to climatic variability over different time scales.

Chapter 3 - A significant fraction of the Near-Earth Object (NEO) population may be objects that originated as comets. Kinetic or nuclear defection schemes may result in the fragmentation of an extinct comet, rather than its deflection. The Solar Collector (SC) has been proposed as a method of deflecting water-ice-rich NEOs on Earth-threatening trajectories. The SC, a two-sail solar sail, functions by concentrating sunlight against the NEO regolith to raise an energized jet that would alter the NEO's solar orbit as a consequence of Newton's Third Law. Recent work reported here expands upon a NASA-supported 2007 study. A dimensionally correct model of SC function is presented with a discussion of energy-loss mechanisms. Measurements of the electromagnetic penetration depth in an extraterrestrial sample are incorporated. The SC-NEO separation is maintained at about 1-km to reduce predicted SC-optics degradation from the energized jet. It is shown that a ~140-m radius SC can divert an ~200-m radius Earth-threatening NEO with a rotational period of about 5 $X10^4$ seconds if the SC remains on station for a period of years. Much smaller ice-rich NEOs in appropriate solar orbits could be diverted into high-Earth-orbits for resource utilization using an SC that remains on station for a sufficiently long time interval.

Chapter 4 - In 2001 during the closest approach to the Earth asteroid 1998 WT24 was observed by Evpatoria–Medicina bistatic radar system. The authors present obtained radar characteristics of this object based on this observation. In this chapter variations in the power spectra of echo signals are analyzed for two orthogonal polarizations of the echo signal. The analysis of the asteroid scattering characteristics for different polarizations indicates stable polarization ratio for the asteroid aspect angles comprising almost the entire asteroid surface. The result of simulation has shown that the scattering law degree of the asteroid roughness is ~1.8; however, the surface is nonuniformly rough. The obtained data also were used for estimation of physical properties of the asteroid showing an average diameter of 405±10 m and large-scale features of the relief on the asteroid surface. The result of measurements of the radar albedo and the surface roughness allow us to present estimations about the properties of the surface material of the asteroid.

Chapter 5 - In this chapter first the statistics of the standard and truncated Pareto distributions are derived and used to fit empirical values of asteroids diameters from different families, namely, Koronis, Eos and Themis, and from the Astorb database. A theoretical analysis is then carried out and two possible physical mechanisms are suggested that account for Pareto tails in distributions of asteroids diameter.

Chapter 6 - Hilda asteroids are objects that orbit the Sun at a distance where the orbital period is exactly 2/3 of the orbital period of Jupiter. Previous analytical and numerical studies have focused their attention to the great dynamical stability within the orbital region where the Hilda asteroids are found. However the group is indeed dispersing. In this paper the authors are going to review the contribution of Hilda asteroids to the Jupiter neiborhood. Based on simulations of long term dynamical evolution of escaped Hilda asteroids, it is find that 8% of the particles leaving the resonance end up impacting Jupiter. Also they are the main source of small craters on the Jovian satellite system, overcoming the production rate by comets and Trojan asteroids. Almost all the escaped Hildas pass through the dynamical region occupied by JFCs, and the mean dynamical life time there is 1.4×10^6 years. ~ 14% of JFCs with $q > 2.5$ A.U. could be escaped Hilda asteroids. We analyzed also the possibility that the Shoemaker - Levy 9 was a Hilda asteroid.

Chapter 7 - The evolution of movement and possible use two asteroids is examined: Apophis and 1950 DA. As a result of the analysis of publications it is established that uncertainty of trajectories of Apophis are caused by imperfection of methods of its determination. The differential equations of motion of Apophis, planets, the Moon and the Sun are integrated by new numerical method and the evolution of the asteroid orbit is investigated. The Apophis will pass by the Earth at a distance of 6.1 its radii on April 13th, 2029. It will be its closest approach with the Earth during next 1000 years. A possibility of transformation of Apophis orbit to an orbit of the Earth's satellite, which can be used for various tasks, is considered. The similar researches have been executed for asteroid 1950 DA. The asteroid will twice approach the Earth to a minimal distance of 2.25 million km, in 2641 and in 2962. It can be made an Earth-bound satellite by increasing its aphelion velocity by ~ 1 km s^{-1} and by decreasing its perihelion velocity by ~ 2.5km s^{-1}.

In: Meteorites and Asteroids
Editors: A. Dementieva et al.
ISBN: 978-1-61942-115-8

Chapter 1

METEORITES AND THEIR ASTEROIDAL PARENT BODIES

Tasha Dunn*
Illinois State University, Normal, IL, US

ABSTRACT

Meteorites are fragments of planetary objects, most often of asteroids, that have broken off of their parent bodies and made their way to Earth. For years, one of the primary goals of planetary scientists who study these objects has been to determine possible connections between recovered meteorites and their potential parent bodies in the asteroid belt and near-Earth asteroid population. Mineralogies and chemistries of meteorites are determined in laboratories, using traditional analytical techniques such as X-ray diffraction and electron microprobe phase analysis. This, along with oxygen isotopic analysis, has resulted in the identification of approximately twenty-five different groups of stony and stony-iron meteorites (those that contain silicate mineral and Fe-Ni metal) and thirteen groups of iron meteorites. Iron and stony-iron meteorites are derived from differentiated asteroids, or those that have a core, mantle, and a crust. Each meteorite group is usually thought to be derived from a different parent body.

The majority of stony meteorites originate from primitive asteroids that did not experience melting, though a smaller number are pieces of crust from differentiated parent bodies. Because of their remote location,

* tldunn@ilstu.edu

asteroids cannot be studied using the analytical techniques applied to meteorites. Instead, they are studied using high-powered telescopes that record the amount of light that is reflected from the asteroid surface. Many minerals preferentially absorb light at particular wavelengths, resulting in characteristic absorption features for these minerals. Visible/near-infrared spectra have been the most widely-applied tool used in the search for these parent asteroids, due to the strong 1 μm and/or 2 μm absorption bands present in the abundant minerals olivine and pyroxene. Abundances and compositions of olivine and pyroxene can be derived by characterizing these spectral features. Comparison of meteorite and asteroid mineralogy has led to linkages between several meteorite groups and possible parent bodies, including two very likely meteorite-asteroid relationships between S-type asteroids and the ordinary chondrites and asteroid 4 Vesta and the HED (howardite, eucrite, and diogenite) meteorites. Though uncertainties in optical, chemical, and physical properties of asteroid surfaces make it difficult to decisively link most asteroids with particular meteorite groups, the search for meteorite parent bodies continues to dominate asteroid spectroscopic studies.

INTRODUCTION

Meteorite Diversity and Classification

As direct samples of the asteroid belt, meteorites can provide insight into the evolution of the early solar nebula, especially the processes by which asteroids formed and were later altered. The more than 52,000 meteorites discovered to date provide ample material for study. Classification of this material provides a mechanism for understanding similarities and differences between the components within, and the processes experienced by, each sample. Because the chemical composition and mineralogy of a sample are functions of the original composition of the parent body, classification also provides a means for determining the number of parent bodies. Most often meteorite parent bodies are asteroids in the main belt, though Mars and the Moon are parent bodies of a small number of Martian and lunar meteorites (~100 distinct falls). Here, we will focus our discussion on meteorites from asteroidal parent bodies.

There are two general types of meteorites, chondritic meteorites and nonchondritic meteorites. Chondritic meteorites, which comprise ~86% of meteorite falls, represent pieces of primitive asteroids that did not experience melting during accretion. They have remained relatively unaltered since the solar system formed. Non-chondritic meteorites are pieces of differentiated asteroids, those that experienced various degrees of partial melting during formation. Differentiated asteroids, like the terrestrial planets, have a core, mantle, and crust. Nonchondritic meteorites include achondrites, primitive achondrites, stony-irons meteorites, and iron meteorites. Each type of nonchondritic meteorite samples a different kind of material (e.g. crust, mantle, or core) within the differentiated parent body.

Chondrites and nonchondritic meteorites are classified based on similarities in bulk chemical composition, mineralogy, petrography (texture), and oxygen isotopic composition. Oxygen isotopes are particularly useful in classification because samples from the same parent body tend to have nearly identical $\delta^{17}O$ and $\delta^{18}O$ compositions (Clayton 1993). Several *classes* of meteorites with similar bulk chemical and isotopic compositions have been established (e.g. Hutchison 2004). Within each class are at least two meteorite *groups* - collections of five or more meteorites with similar mineralogy and petrography. Smaller collections, with only two to four samples, are called *grouplets*, and meteorites that differ from any known group or grouplet are referred to as *ungrouped* or *anomalous*. Mineralogy and bulk chemistry of meteorites are analyzed in laboratories, using traditional analytical techniques such as X-ray diffraction and electron microprobe phase analysis. This, along with oxygen isotopic analysis, has resulted in the identification of thirteen groups of chondrites and two chondrite grouplets, fifteen groups of achondrite and primitive achondrites, four groups of stony-irons, and thirteen groups of iron meteorites. Though there are numerous ungrouped chondrites and achondrites, the iron meteorites have the most ungrouped samples, with more than 10% of iron meteorites (more than 100 samples) falling into this category.

Classification of Chondrites

All chondrites have solar-like compositions (with the exception of highly volatile elements), with the CI chondrites representing the best estimate of solar compositions (Anders & Grevesse 1989). Many (but not all) chondrites contain chondrules, small (1-2 mm) spheres of molten material from which they are named, refractory inclusions, and metal. There are three major classes of chondritic meteorites: carbonaceous chondrites, ordinary chondrites, and enstatite chondrites. Ordinary chondrites are named as such because they are the most common type of chondrite and enstatite chondrites because they consist primarily of the pyroxene mineral enstatite. The term carbonaceous, however, is a historical misnomer, as we now know that most C chondrites do not contain appreciable amounts of carbon. The most recent classification of chondritic meteorites is shown in Fig. 1.

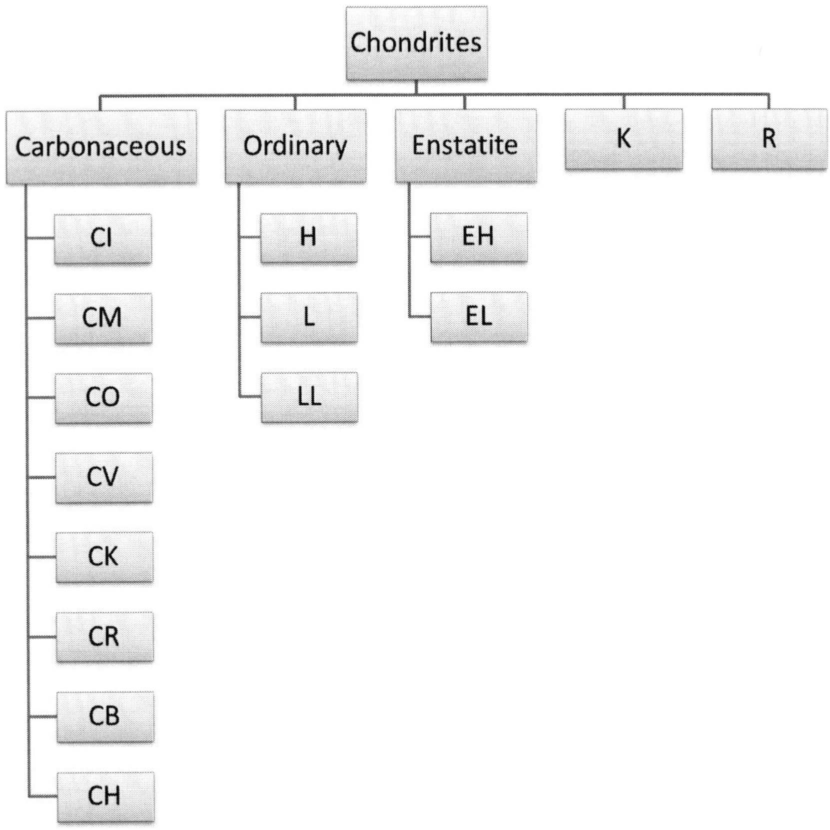

Figure 1. Classification of nonchondritic meteorites showing the three chondrite classes (carbonaceous, ordinary, and enstatite), the many groups, and the K and R grouplets.

The three chondrite classes are divided into several groups, each of which has similar petrologic, chemical, and isotopic characteristics. The carbonaceous chondrite class consists of eight groups (CI, CM, CO, CV, CK, CR, CB, and CH). Each carbonaceous chondrite group is identified by the letter C followed by an additional letter indicating the type specimen for that group (e.g. the CI chondrites are named after the Ivuna meteorite, which landed near Ivuna, Tanzania on December 16, 1938). The ordinary chondrites are divided into three groups (H, L, and LL) based on the abundance of Fe-Ni metal. The H chondrites contain the highest abundance of metal and the LL chondrites the lowest. The enstatite chondrite class consists of only two groups, the EH (high Fe) and EL (low Fe) groups. Two groups of chondrites, the R and the K chondrites, do not fall within any of the three major classes (Fig 1). The R (Rumurutti-like) chondrites have compositions similar to the ordinary chondrites, yet their oxygen isotope compositions differ, suggesting that they may have originated from a different parent body. The K (Kakangari-like) chondrites are a group of three chondrites that share some characteristics of each of the three chondrite classes, but together their properties plot outside those the major chondrite classes.

In addition to the chemical classification scheme described above, chondrites are also classified based petrologic type (Van Schmus & Wood 1967; McSween et al. 1988). In this classification scheme, chondrites are assigned a number (1-6) corresponding to the degree of thermal metamorphism or aqueous alteration that the sample experienced. Chondrites that are relatively unmetamorphosed, or have experienced very slight heating, are designated as type 3 (unequilibrated), whereas chondrites that have undergone the highest degrees of thermal metamorphism are designated as type 6 (equilibrated) (Van Schmus and Wood 1967; McSween et al. 1988). Petrologic types 2 and 1 represent increasing degrees of aqueous alteration, which is most prevalent in the CI, CM, and CR chondrites (McSween 1979). Most chondrite groups have experienced some degree of metamorphism. In the ordinary chondrites and CO chondrites, metamorphic effects are particularly prominent in petrologic type 3 material. As a result, type 3 ordinary and CO chondrites have been subdivided into petrologic subtypes (i.e. 3.0 – 3.9) (Sears et al. 1980; Sears et al. 1991; Chizmadia et al. 2002).

Classification of Nonchondritic Meteorites

Nonchondritic meteorites represent pieces of asteroids that have melted and therefore experienced some degree of igneous processing (i.e. partial melting or crystal fractionation). They are much more diverse than their primitive counterparts, with fifteen groups of primitive and differentiated achondrites, four groups of stony-iron meteorites, and thirteen groups of iron meteorites currently recognized. Achondrites, like chondrites, are stony meteorites, though they lack the primitive components, such a chondrules and refractory inclusions, present in chondrites. Stony-iron meteorites contain roughly equal amounts of silicate and metal and are believed to represent material from the core and mantle or crust of a single asteroid. Iron meteorites, composed entirely of Fe-Ni metal, are remnants of asteroid cores. The classification for differentiated meteorites is shown on Fig. 2.

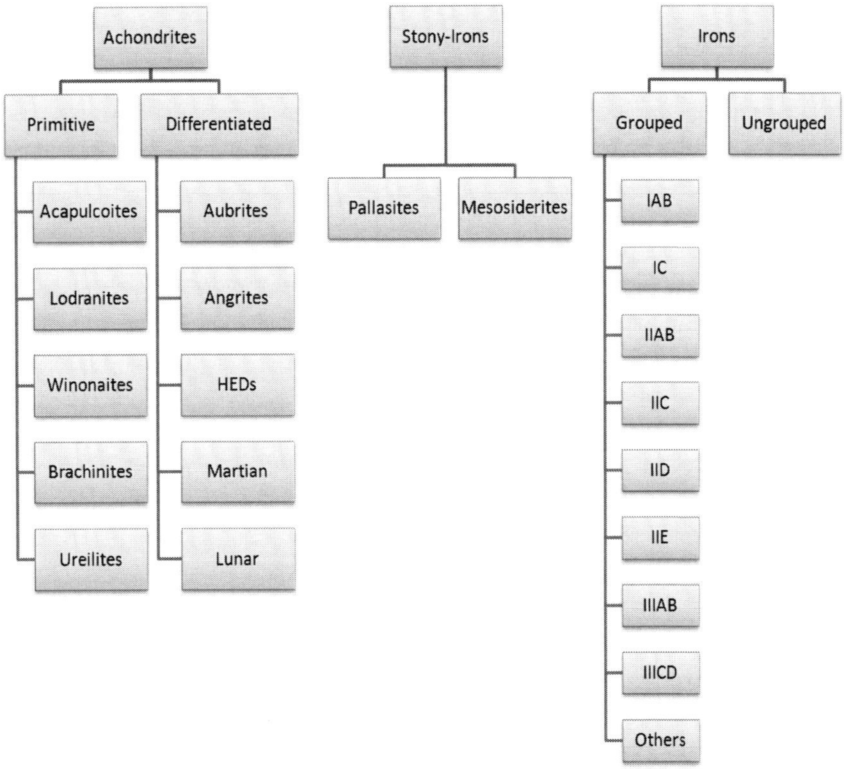

Figure 2. Classification of the three different types of nonchondritic meteorites: achondrites, stony-irons, and irons. The box labeled HEDs is actually three genetically-related groups: the howardites, eucrites, and diogenites.

Primitive achondrites. Though all achondrites have igneous textures, some retain chondritic compositions. These achondrites are called the primitive achondrites. Primitive achondrites are the residual solid material left over after partial melting has occurred. They retain their chondritic compositions because they form as the result of only a small degree of partial melting. Five groups of achondrites are designated as primitive: acapulcoites, lodranites, brachinites, winonaites, and ureilites. Acapulcoites and lodranites are similar to one another in composition and texture and are often referred to as the acapulcoite-lodranite clan (Weisberg et al. 2006). Meteorites in both groups have mineralogies similar to the ordinary chondrites, though plagioclase and troilite as less abundant in the lodranites (McCoy et al. 1996; 1997; Nagahara 1992; Mittlefehldt et al. 1996). Acapulcoites likely formed from low degrees of partial melting, while lodranites are thought to be residues from higher degrees of partial melting (Weisberg et al. 2006). Winonaites are distinguished from acapulcoites and lodranites based on oxygen isotopes (Clayton and Mayeda 1996). They have compositions similar to ordinary and enstatite chondrites and sometimes contain relict chondrules. Fe-Ni-sulfide veins present in many winonaites are thought to represent the first partial melts of a chondritic precursor (Benedix et al. 1998).

Brachinites are small group of achondrites consisting primarily of Mg-rich olivine ($Fo_{65\text{-}70}$) (Warren and Kallemeyn 1989). Their origin is not as clear as the other primitive achondrites. It has been suggested that they are recrystallized chondritic material (Warren and Kallemeyn 1989) or partial melt residues (Nehru et al. 1996). However, Mittlefehldt (2004) proposed that the brachinites were igneous cumulates, which would indicate a magmatic (differentiated) origin. The ureilites are the largest group of primitive achondrites. They consist primarily of olivine and pyroxene, similar to mantle peridotites, but with some carbon in the form of graphite and microdiamonds. Because there are three different types of ureilites, their origin is even more perplexing than the brachinites. It is thought that some ureilites are partial melt residues (e.g. Warren and Kallemeyn 1992; Goodrich et al. 2002) while others may be cumulates (Goodrich et al. 1987; 2001).

Differentiated achondrites. The igneous textures and non-chondritic compositions of the differentiated (magmatic) achondrites are the result of complete melting of chondritic precursor material. Because partial melting of chondritic material yields basaltic magmas, differentiated achondrites are always basalts or cumulates derived from basaltic magma. Chemical and mineralogical differences between the achondrite groups are the results of slight variations in precursor material. Differentiated achondrites include meteorites from the moon and Mars, though three groups of differentiated achondrites originate from asteroids. There are aubrites, angrites, and the HED clan (howardites-eucrites-and diogenites).

Aubrites are often referred to as enstatite achondrites because of their similarities to the enstatite chondrites, though they are not thought to be derived from the same parent body (Keil 1989). The highly-reduced aubrites are comprised mostly of FeO-free enstatite with minor amounts of plagioclase,

FeO-poor diopside, and olivine (Watters and Prinz 1979). Angrites are basalts with unusual mineralogies of Al- and Ti-rich diopside and Ca-rich olivine in addition to silica-poor minerals such as nepheline (e.g. Mittlefehldt and Lindstrom 1990; Mittlefehldt et al. 1988). They are thought to have formed as partial melts of a CAI-rich precursor under oxidizing conditions (Prinz et al. 1977). The howardite-eucrite-diogenite (HED) clan is the largest group of differentiated meteorites. Diogenites are cumulates consisting mostly of orthopyroxene and small amounts of olivine and chromite. Two types of eucrites have been identified: basaltic eucrites and cumulate eucrites. Both are similar in composition, consisting primarily of pigeonite (pyroxene) and plagioclase, but differ in texture. Basaltic eucrites are fine-grained, suggesting formation by rapid cooling of magma, while cumulate eucrites cooled at depth and are coarser-grained. Howardites are impact breccias containing pieces of eucrite and diogenite material (Fredriksson and Keil 1963; Delaney et al. 1984).

Stony-irons. There are only two groups of stony-iron meteorites, the pallasites and the mesosiderites. Both contain roughly equal amounts of metal and silicate minerals. The difference between the two groups is the type of silicate minerals present. Pallasites contain olivine (a mantle-derive silicate), while mesosiderites contain silicate minerals from the crust, primarily pyroxene. Pallasites represent material from the core-mantle boundary, while mesosiderites are thought to be pieces of material from the crust and the core that were mixed during an impact.

Irons. Thirteen groups of iron meteorites have been identified, though at least 10% of iron meteorites cannot be placed into one of the established groups. All iron meteorites are composed almost entirely of Fe-Ni metal, which typically forms as intergrowths of high-Ni and low-Ni metal (Haack and McCoy 2004). The classification of the iron meteorites is perhaps the most convoluted classification scheme of all known meteorite types, primarily because of its historical development. Initially, four groups of irons were established based on Ga and Ge concentrations. These groups were labeled using the Roman numerals I-IV (Lovering et al. 1957). As more samples were analyzed, subgroups within each group were recognized. These new groups were distinguished by adding letters to the Roman numeral (i.e. group II became IIA and IIB). Over time, several groups have been combined, forming groups such as the IIAB and IIIAB irons.

Measuring and Interpreting Asteroid Spectra

Asteroids are studied using reflectance spectroscopy, a method in which the amount of incident sunlight that is reflected or absorbed by the surface of an asteroid is measured by Earth-bound telescopes. The amount of light that is reflected and absorbed is a function of a number of surficial properties, including composition, particle size, and temperature. All minerals absorb light at certain wavelengths, producing characteristic absorption features. Many silicate minerals common in meteorites (e.g., olivine, pyroxene) have characteristic absorption features in the visible and near-infrared wavelengths (VISNIR), making them particularly useful for asteroid spectroscopy since the atmosphere is relatively transparent in this wavelength region and the Sun's blackbody curve peaks in the visible. The NASA Infrared Telescope Facility (IRTF) SpeX instrument (Rayner et al. 2003) located on Mauna Kea observes objects in the near-infrared (0.8-2.5 μm). SpeX data is combined with visible spectra (0.4 – 1.1 μm) from surveys such as SMASS (small-main-belt asteroid survey) (Xu et al. 1995; Bus and Binzel 2002a, 2002b) or Palomar (Binzel et al. 2001) to allow for full wavelength coverage between ~0.4 and ~2.5 μm.

Visible Analysis of Spectra

In the visible wavelengths (0.4 – 0.9 μm), two features are commonly used to describe and classify a spectrum: the presence or absence of an absorption feature that extends from the ultraviolet to ~0.7 μm and the presence or absence of a silicate absorption feature at ~1 μm. Asteroids were historically classified based on their spectral properties in the visible wavelength region but as near-infrared spectra became easier to obtain, this infrared data have been incorporated into taxonomies. The first classification for asteroids based on spectral reflectance in the visible wavelength range was proposed by Chapman et al. (1975). Since then several taxonomies have been developed (Tholen 1984; Barucci et al. 1987; Tedesco et al. 1989; Howell et al. 1994), all of which yield fundamentally similar results. The most commonly used taxonomies are the Tholen taxonomy (Tholen 1984) and the SMASSII taxonomy (Bus & Binzel 2002b). The Tholen taxonomy consists of 14 classes of asteroids, the most common of which being the A-types, C-types, S-types, V-types, and X-types. Using visual albedo, the X-types are broken into the E-types (high albedo), M-types (moderate albedo), and P (low albedo). Based on the analysis of CCD (charge-coupled device) spectra, the SMASSII taxonomy builds on the Tholen taxonomy by subdividing many Tholen classes into smaller groups, increasing the number of asteroid classes from 14 to 26 due to the diversity of the CCD spectra of asteroids. The DeMeo et al. (2009) taxonomy classifies asteroids with visible and near-infrared spectra into 24 classes using the SMASSII taxonomy as a guide. Though taxonomy is useful for grouping objects with similar spectral shapes, it is important to note that it is not always diagnostic of mineralogy or composition. Visible spectra must be combined with near-infrared spectra in order to determine mineralogy.

Visible/Near-Infrared Mineralogical Analysis of Spectra

Visible/near-infrared spectra have been widely used to determine mafic mineral abundances and compositions of asteroids since the relationship between spectral properties (i.e. absorption features) and mineralogy in meteorites was first recognized (Adams 1974; Adams 1975; Burns 1970; Cloutis 1985; Cloutis and Gaffey 1991). Most previous work examining the spectral properties of silicate minerals has focused on olivine and pyroxene, both of which have diagnostic spectral properties in the VISNIR. Olivine and pyroxene are also the two primary components of ordinary chondrites, and are found extensively in many types of non-chondritic meteorites. The presence of olivine and pyroxene can usually be easily identified in asteroid spectra due to the distinctive absorption bands of olivine and pyroxene.

The primary diagnostic feature in olivine is a composite absorption feature at ~1 μm (Fig. 3), which consists of three distinct absorption bands at roughly 0.9 μm, 1.1 μm, and 1.25 μm. The composite 1 μm band, which is attributed to electronic transitions of Fe^{2+} that occupy both the M1 and M2 crystallographic sites (Burns 1970), moves to longer wavelengths as FeO content increases (King and Ridley 1987; Sunshine and Pieters 1998). Pyroxenes have two absorption bands, one at ~1 μm (Band I) and one at ~2 μm (Band II) (Fig. 3), which are associated with crystal field transitions in Fe^{2+} that preferentially occupy the M2 site (Clark 1957; Burns 1970). Low-calcium pyroxenes, which are conventionally defined as having less than 11 mol% $CaSiO_3$ (wollastonite or Wo) (Adams 1974), show a well-defined relationship between absorption band positions and composition, as both Band I and Band II positions move to longer wavelengths with increasing ferrous iron content (Adams 1974; Cloutis 1985). There is also a correlation between composition and band positions in high-calcium pyroxene, although the relationship is complicated by the presence of calcium in addition to iron. The ratio of the area of Band II to

Band I is called the Band Area Ratio (BAR) and can be used to distinguish pyroxene-rich bodies (high BAR) and olivine-rich bodies (low BAR).

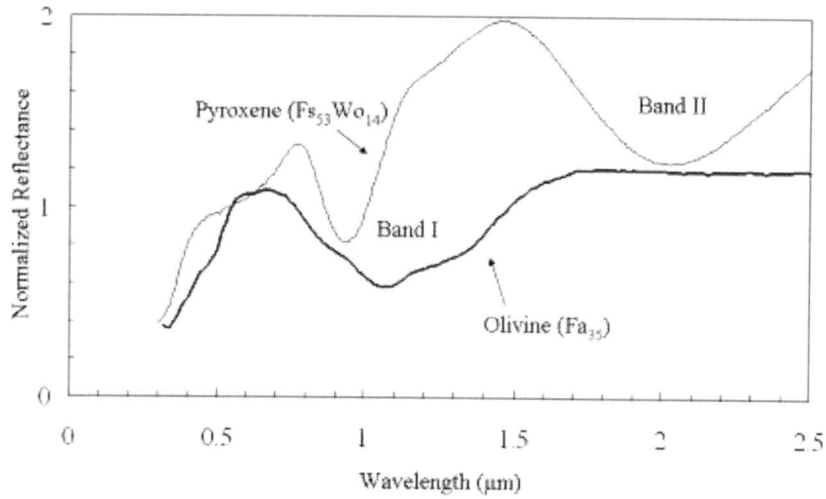

Figure 3. Reflectance spectra of olivine and pyroxene. Olivine spectrum is for EET 99402 (Fa35) (brachinite) (Burbine et al. 2006). Pyroxene spectrum is for Bouvante (eucrite) (Fs53Wo14) (Burbine et al. 2001a). Spectra are normalized to unity at 0.55 μm.

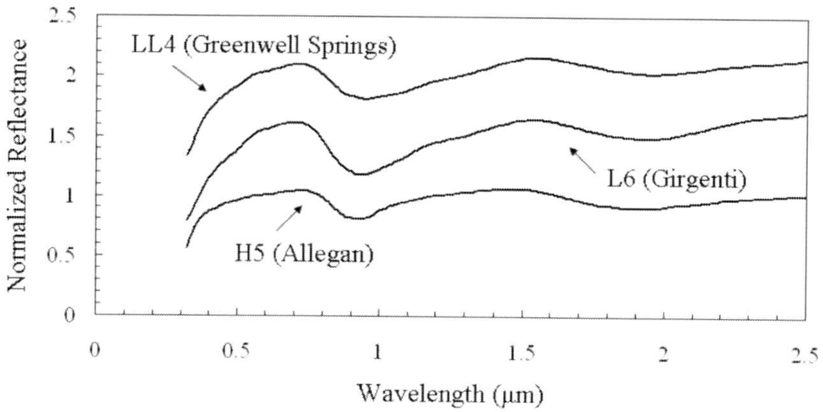

Figure 4. Reflectance spectra of an H-, L-, and LL-chondrite. Data from Burbine et al. (2003). The band positions (Band I and Band II) move to longer wavelengths as you go from an H to an L to an LL chondrite. The LL chondrites also have the most distinctive absorption bands due to olivine. Spectra are offset in reflectance.

Pyroxene grains tend to spectrally dominate any olivine-pyroxene mixtures in the visible and near-infrared since pyroxene is more absorbing than olivine in this wavelength region. These effects can be seen in the spectral properties of ordinary chondrites (Fig. 4). H chondrites, which contain the lowest ratio of olivine to pyroxene, have the strongest pyroxene absorption features, whereas LL chondrites, which contain the highest ratio of olivine to pyroxene, have the strongest olivine absorptions. L chondrites are intermediate between H and LL chondrites. The Band II positions for LL chondrites tend to be at longer wavelengths than L chondrites because their pyroxenes tend to be richer in Fe. Further, the Band II positions for L chondrites tend to be at longer wavelengths than H chondrites for the same reason.

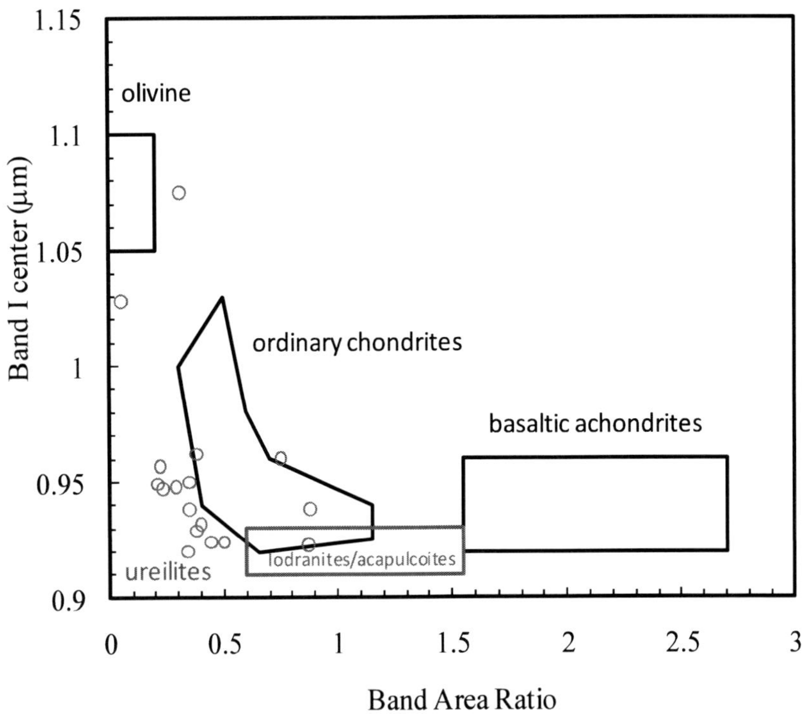

Figure 5. Olivine, ordinary chondrite, and basaltic achondrite regions on a plot of Band Area Ratio versus Band I center (μm) (Gaffey et al. 1993). The primitive achondrite region (red box) is from Burbine et al. (2001a). The ureilite values (blue circles) are from Cloutis et al. (2010).

There are also a few primitive achondrites (e.g., lodranites/acapulcoites, ureilites) that also have olivine and pyroxene bands. Lodranites/acapulcoites are primarily mixtures of low-Ca pyroxene ($Fs_{6.5-12.6}$) and olivine (Fa_{5-12}) that have experienced partial melting (e.g., McCoy et al. 2000). Spectral measurements of these meteorites (Hiroi and Takeda 1991; Burbine et al. 2001b) find that their spectra are dominated by low-Ca pyroxene. In a plot of Band Area Ratio versus Band I center (Figure 5), the lodranites/acapulcoites either fall between the ordinary chondrite and basaltic achondrite (howardites, eucrites, diogenites) regions or overlap the most pyroxene-rich ordinary chondrites (H chondrites). Besides olivine and pyroxene, ureilites contain 10% or less dark interstitial material that consists of varying amounts of carbon, metal, and sulfides (Mittlefehldt et al. 1998). Ureilites have pyroxene and olivine absorption features that are subdued in strength (Cloutis et al. 2010) relative to ordinary chondrites due to their higher carbon contents. In a plot of Band Area Ratio versus Band I center (Fig. 5), the ureilites tend to have Band Area Ratios slightly lower than those found for ordinary chondrites with the same Band I Centers as the ureilites. However, a few ureilites do fall in the ordinary chondrite region. F-type 2008 TC3 that exploded in the atmosphere and rained fragments over the Sudan was found to have a ureilite composition from analyses of the recovered samples (Jenniskens et al. 2009).

Mafic Mineral Abundances. BAR is particularly useful because it is sensitive to the relative proportions of olivine and pyroxene. BAR (Cloutis et al. 1986) is commonly used to estimate olivine and pyroxene abundances in meteorites and asteroids. The band area ratio (BAR) can be expressed a linear regression function relating band area ratio to relative abundances of olivine and pyroxene:

$$px/(ol+px) = (0.417 \times BAR) + 0.052$$ (Cloutis et al. 1986).

Because the Cloutis et al. (1986) regression was based on simple mixtures of olivine and orthopyroxene, the presence of more than one pyroxene (or other additional phases) could complicate spectral interpretations made using this calibration. To account for the presence of more than one pyroxene, Burbine et al. (2003) revised the original Cloutis et al. (1986) regression using normative abundances of olivine, orthopyroxene, and clinopyroxene in ordinary chondrites. Burbine et al. (2003) expressed their equation as:

$$ol/(ol+px) = -0.228 \times BAR + 0.768.$$

Most recently, Dunn et al. (2010c) revised the BAR linear regression, this time using measured mineral abundances and accounting for pigeonite, the third pyroxene present in the ordinary chondrites. Because this regression was derived using measured abundances (as opposed to abundances calculated using CIPW norms), it provides a more accurate measure of olivine and pyroxene abundances from VISNIR spectra. Dunn et al. (2010c) expressed their equation as:

$$ol/(ol+px) = -0.242 \times BAR + 0.728$$

The Dunn et al. (2010c) regression formula is well suited for asteroids with ordinary chondrite-like compositions, but Cloutis et al. (1986) is better suited for asteroids that are compositionally similar to lodranites and acapulcoites, which are primarily mixtures of low-Ca pyroxene and olivine (e.g., McCoy et al. 2000). Because different formulas work better for different assemblages, caution must be used when applying them to any asteroid spectrum.

Mafic Mineral Compositions. Compositions of mafic minerals can be determined from their Band I and Band II centers. Gaffey et al. (2002) devised the following series of equations defining the relationships between absorption-band centers and pyroxene composition, expressed as pyroxene endmembers wollastonite (Wo) and ferrosilite (Fs), using the pyroxene data of Cloutis and Gaffey (1991). The values in parentheses are the compositional ranges for which each equation is valid.

Wo (±3) = 456.2 x BI Center (μm) – 416.9 (Fs = 10-25; Wo10-35 excluded)

Fs (±5) = 57.5 x BII Center (μm) – 72.7 (Wo = 11-30; Fs<25 excluded)

Fs (±5) = 268.2 x BII Center (μm) – 483.7 (Wo < 11)

Wo (±4) = 418.9 x BI Center (μm) – 380.9 (Fs = 25-50)

Fs (±5) = 268.2 x BII Center (μm) – 483.7 (Wo = 30-45)

Fs (±4) = -118.0 x BII Center (μm) + 278.5 (Wo > 45)

Burbine et al. (2007) developed a second set of equations for deriving pyroxene chemistry from band centers using a suite of thirteen HED meteorites. The Burbine et al. (2007) equations are simpler to use because they work only on the very restricted range of pyroxene compositions found in the HEDs. These equations are:

Fs (±3) = 1023.4 x Band I center – 913.82

Fs (±3) = 205.86 x Band II center – 364.3

Wo (±1) = 396.13 x Band I center – 360.55

Wo (±1) = 79.905 x band II center – 148.3

Because the Gaffey et al. (2002) equations tend to overestimate the Fs content in ordinary chondrite-like assemblages, Dunn et al. (2010c) used ordinary chondrite powders originally prepared by Jarosewich (1990) and mineralogically and spectrally characterized by Dunn et al. (2010a, 2010b) to develop two additional formulas for determining the mineralogies of ordinary chondrites and their parent asteroids from reflectance spectra. The derived formulas for determining the mineralogies of ordinary chondrite assemblages are:

Fa (±1.3) (mol%) = -1284.9*(Band I center)^2 + 2656.5 x (Band I center) – 1342.3

Fs (±1.4) (mol%) = -879.1*(Band I center)^2 + 1824.9 x (Band I center) – 921.7

As mentioned earlier, caution must be used when applying any of these formulas for deriving mineralogy to an asteroid spectrum since different formulas work better for different assemblages. Coupled with the fact that these formulas have not been fully tested on returned asteroidal samples, except for the returned fragments from the Hayabusa mission to asteroid 25143 Itokawa, how well these formulas predict asteroid mineralogies is still up for debate.

Effects of Surface Properties on Asteroid Spectra

When comparing asteroid spectra to meteorite spectra, it is important to keep in mind that the surfaces of asteroids are subject to processes that meteorites are not, such as space weathering, and that other factors, such as temperature, can alter the spectrum of an object. We must understand how these factors affect the optical properties of an object before we can make interpretations about potential relationships between asteroids and meteorites. There are four primary factors that can alter asteroid spectra: space weathering, phase reddening, particle size, and temperature.

Space Weathering. Space weathering (e.g., Hapke 2001) is the term for the processes that can potentially alter the spectral properties of the surface of an "airless" body. These processes include the interaction with galactic and solar cosmic rays, irradiation by solar wind particles, and micrometeorite and meteorite impacts. These processes darken the surface, redden the spectral slope, and reduce the strength of absorption bands through the production of nanophase iron (e.g., Sasaki et al. 2001). By analyzing the spectral slope of members of asteroid families, Vernazza et al. (2009) argues that solar wind irradiation rapidly weathers the surfaces of asteroids on timescales of 10^4-10^6 years (Strazulla et al. 2005). Gaffey (2010) argues that space weathering does not affect spectral parameters such as Band Centers and Band Area Ratios, which are important for determining mineralogy. Binzel et al. (2010) found that near-Earth asteroids classified as Q-types have experienced close approaches to Earth (passed within the Earth-Moon distance) within the last ~0.5 million years and argues that seismic shaking during close encounters with Earth exposes unweathered surface material and changes the spectral properties of asteroids with ordinary chondrite compositions so that the objects appear less space weathered.

Phase Reddening. Phase reddening is the increase in the spectral slope of an asteroid with increasing phase angle (Sun-asteroid-Earth angle). This reddening has been seen in the laboratory (Gradie et al. 1980) and spacecraft observations of Eros (Clark et al. 2002) and Itokawa (Abe et al. 2006). The origin of phase reddening is not well understood (e.g., Muinonen et al. 2002).

Particle Size. Particle size can also affect the spectral properties of an asteroid. Decreasing the particle size of a sample tends to increase the spectral slope and decrease the strength of any absorption bands. The actual particle sizes on asteroids are unknown but the best asteroid spectral matches to meteorite spectra tend to be for meteorite samples sieved to relatively small sizes.

Temperature. Though laboratory-acquired meteorite spectra are obtained at room temperature, asteroid surface temperatures range from 120K to 300K (Hinrichs et al. 1999). The band positions of pyroxenes move to shorter wavelengths for lower surface temperatures (Singer and Roush 1985; Roush and Singer 1987; Hinrichs et al. 1999; Moroz et al. 2000; Burbine et al. 2001a). Moroz et al. (2000) found that the Band II Centers always move to shorter wavelengths with decreasing temperatures in olivine-pyroxene assemblages while the direction of movement of the Band I center depends on the percentage of olivine in the sample. Band Area Ratios for olivine-pyroxene assemblages tend to increase with decreasing temperature (Moroz et al. 2000). Asteroid surface temperatures can be estimated and then corrected for using the asteroid albedo, the solar luminosity, the beaming factor (usually assumed to be unity), the asteroid's infrared emissivity, and the asteroid's distance from the Sun. At temperatures ~200-300 K, the correction due to temperature is relatively small (<0.01 μm for Band I and <0.02 μm for Band II) (Moroz et al. 2000; Burbine et al. 2001a).

METEORITES AND POSSIBLE PARENT BODIES

The exact number of asteroidal parent bodies represented in the meteorite collection is unclear, but meteorite classifications suggest that there are likely as many as 100-200 distinct parent bodies. This number of postulated parent bodies is extremely small compared to the number of known main-belt asteroids (~600,000) and bodies larger than 1 km in diameter that are thought to exist in the belt (~1.9×10^6) (Tedesco et al. 2005). Twenty seven parent bodies are required for the thirteen chondrite groups and two grouplets, though the CV chondrites may represent as many as three parent bodies (McSween 1977). There are at least six parent bodies for the differentiated achondrites, four for the stony irons, and 60 for the grouped and ungrouped irons (Burbine et al. 2002a). It is possible that this number may be even higher if potential relationships between the ungrouped irons are found to be absent.

Ordinary Chondrites and the S-Type Asteroids

Since the first comprehensive study of ordinary chondrite spectra was completed (Gaffey 1976), the ongoing search for the parental asteroids of the ordinary chondrites has centered on the S(IV) subgroup, one of seven subgroups of the S-type asteroids (Gaffey et al. 1993; Gaffey and Gilbert 1998).

Table 1. Meteorite groups and their postulated parent or souce bodies*

Group	Postulated Parent or Source Bodies†
L	S(IV) asteroids (Gaffey et al., 1993)
H	6 Hebe [S(IV)] (Gaffey and Gilbert, 1998)
LL	S(IV) asteroids (Gaffey et al., 1993)
R	A or S asteroids
EH	M asteroids (Gaffey and McCord, 1978)
EL	M asteroids (Gaffey and McCord, 1978)
CV	K asteroids (Bell, 1988)
CI	C asteroids (Gaffey and McCord, 1978)
CO	221 Eos (K) (Bell, 1988)
CM	19 Fortuna (G, Ch) (Burbine, 1998)
CR	C asteroids (Hiroi et al. 1996)
CK	C asteroids (Gaffey and McCord, 1978)
CH	C or M asteroids
(Tagish Lake)‡	D asteroids (Hiroi et al. 2001)
Eucrites	4 Vesta (V) (Consolmagno and Drake, 1977; Drake, 2001)
Howardites	4 Vesta (V) (Consolmagno and Drake, 1977; Drake, 2001)
Diogenites	4 Vesta (V) (Consolmagno and Drake, 1977; Drake, 2001)
Mesosiderites	M asteroids (Gaffey et al. 1993)
Pallasites	A asteroids (Cruikshank and Hartmann, 1984; Lucey et al. 1998)
Ureilites	S asteroids (Gaffey et al. 1993); F types (Jenniskens et al. 2009)
Aubrites	3103 Eger (E) (Gaffey et al. 1992)
Acapulcoites	S asteroids (McCoy et al. 2000)
Angrites	S asteroids (Burbine et al. 2001a)
Lodranites	S asteroids (Gaffey et al. 1993; McCoy et al. 2000)
Winonaites	S asteroids (Gaffey et al. 1993)
Brachinites	A asteroids (Cruikshank and Hartmann, 1984; Sunshine et al. 1998)
Irons	M asteroids (Cloutis et al. 1990; Magri et al. 1999)
"Martian"	Mars (McSween, 1994)
"Lunar"	Moon (Warren, 1994)

*Modified from Burbine et al. (2002)

†Asteroid classes are a combination of those of Tholen (1984), Gaffey et al. (1993), and Bus (1999).

‡Tagish Lake is an unusual carbonaceous chondrite (Brown et al., 2000) and is listed in the table because of its spectral similarity to D asteroids.

The S-type asteroids represent a suite of mixtures ranging in composition from pure olivine to pure pyroxene (with potential meteorite analogues including ureilites [S(II)], lodranites [S(III) and S(V)] and mesosiderites [S(VII)] (Gaffey et al. 1993) (Fig. 6). The S(IV) subgroup is thought to contain objects with mineralogies similar to the ordinary chondrites (Gaffey et al. 1993). The S(IV) asteroid 6 Hebe has been hypothesized to be the parent body of the H chondrites (Gaffey and Gilbert 1998), and the Flora Family has been suggested as possible source of the L chondrites (Nesvornỳ et al. 2002). In addition, a large portion of studied Near Earth Asteroids (NEAs) has reflectance spectra similar to ordinary chondrites (Thomas and Binzel 2009). Like the S(IV) asteroids, ordinary chondrites scatter along a mixing line between olivine and low-Ca pyroxene when plotted in band area ratio vs. Band I center space (Fig. 2). This representation suggests that ordinary chondrites form a nearly continuous sequence between olivine-rich LL chondrites and relatively pyroxene-rich H chondrites (Gaffey et al. 1993).

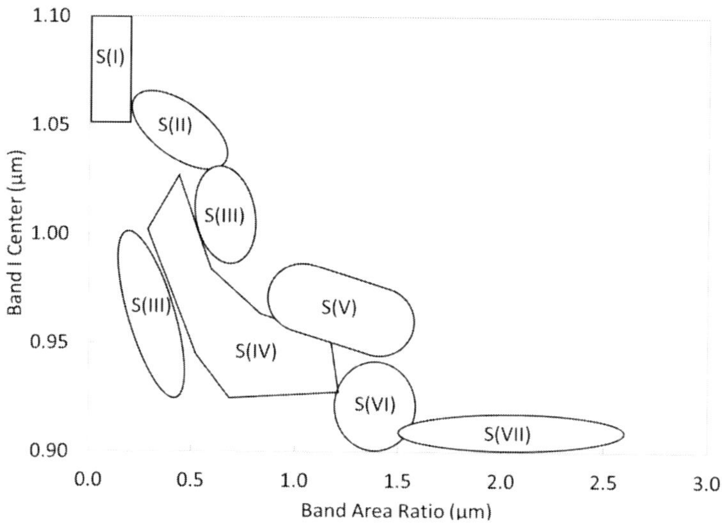

Figure 6. Spectral classification of S-type asteroids into seven spectral classes (modified from Gaffey et al. (1993) and Gaffey and Gilbert (1998)).

Spacecraft missions appear to "confirm" a relationship between some S-types and ordinary chondrites and that space weathering occurs on these objects. NEAR Shoemaker orbited and then landed on near-Earth S-asteroid Eros from February of 2000 to February of 2001. The surface mineralogy of Eros was estimated using reflectance spectra while its surface chemistry was estimated using the first X-ray and gamma-ray measurements of an asteroid. Spectral properties (e.g., Band Area Ratios) derived from reflectance measurements were consistent with ordinary chondrites. The newest (Foley et al. 2006; Lim and Nittler 2009) calibrated Mg/Si, Al/Si, Ca/Si, Fe/Si, Cr/Fe, Mn/Fe, and Ni/Fe elemental abundance ratios that were derived from X-ray measurements of Eros are consistent with ordinary chondrites though S/Si is depleted. This sulfur depletion is believed to be due to space weathering where troilite is vaporized and sulfur is lost (Trombka et al. 2000). Looking at all available data, McCoy et al. (2001) found that the mineralogy of Eros was best matched by an ordinary chondrite that was altered at the surface of the asteroid or a primitive achondrite derived from material mineralogically similar to ordinary chondrites.

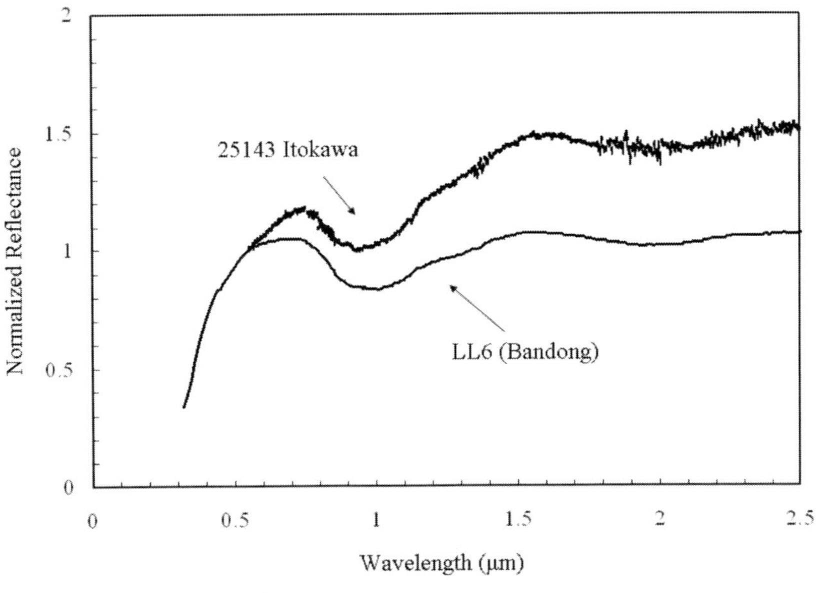

Figure 7. Reflectance spectra of S-type NEA 25143 Itokawa (Binzel et al. 2001) and LL6 chondrite Bandong (Burbine et al. 2003). Spectra are normalized to unity at 0.55 μm. Error bars are one sigma.

Hayabusa was the first spacecraft to return samples of an asteroid back to Earth. Hayabusa rendezvoused with S-asteroid 25143 Itokawa in September of 2005, landed briefly on the object to obtain samples, and then returned back to Earth where it landed in Australia in June of 2010. Prior to the Hayabusa return mission, there were contradictory interpretations of Itokowa's mineralogy. Using visible and near-infrared spectra of Itokawa (Fig. 3) taken from Earth, Binzel et al. (2001) suggested the surface mineralogy of Itokawa was analogous to a space-weathered LL chondrite (Fig. 7). Binzel et al. (2001) made their interpretation using a simple space weathering model and MGM (Modified Gaussian Modeling). Abell et al. (2007) used BAR formulas from Gaffey et al. (2002) to suggest that the surface mineralogy of Itokowa was consistent with a partially melted body. This scientific disagreement was "solved" when laboratory analyses of fragments of Itokawa showed that the returned samples of Itokawa were consistent with LL chondrites (Nakamura et al. 2011).

HEDs and Asteroid 4-Vesta

With a diameter of ~500 km, Vesta is the third largest asteroid and the largest member of the Vesta family of asteroids. Asteroid 4 Vesta has been linked to the HED meteorites since the first visible spectrum of Vesta was obtained (McCord et al. 1970). The visible spectrum of Vesta was similar to the distinctive spectra of the pyroxene-rich basaltic HEDs (Fig. 8). Larson and Fink (1975) and McFadden et al. (1977) confirmed this spectral similarity with near-infrared spectral observations, and Hiroi et al. (1994) found the best spectral match to Vesta is a fine-grained (< 25 μm) howardite. Binzel and Xu (1993) found that asteroids in the Vesta family located between Vesta and the 3:1 and ν_6 meteorite-supplying resonances also had this distinctive HED-like spectrum.

However, Vesta is not the only body that formed in the asteroid belt with a basaltic crust. Asteroid 1459 Magnya (Lazzaro et al. 2000; Hardersen et al. 2004) was observed to have an HED-like spectrum. Magnya is located in the outer main belt at 3.14 AU in a location where it is extremely difficult to derive from Vesta, which is located at 2.36 AU. A number of objects with HED-like spectra have been observed (e.g., Roig et al. 2008) in the middle and outer main-belt far from Vesta.

CM Chondrites and the C-Type Asteroids

CM chondrites tend to have a distinctive absorption feature at 0.7 μm (Vilas and Gaffey 1989; Hiroi et al. 1993) that is not found in the spectra of other carbonaceous chondrites. This is a charge transfer absorption between

Fe^{2+} and Fe^{3+} in phyllosilicates (Vila and Gaffey 1989). This feature disappears when CM material is heated to temperatures above 400 °C (Hiroi et al. 1993). CM chondrites also have a strong feature at ~3 μm due to H_2O or OH in hydrated silicates.

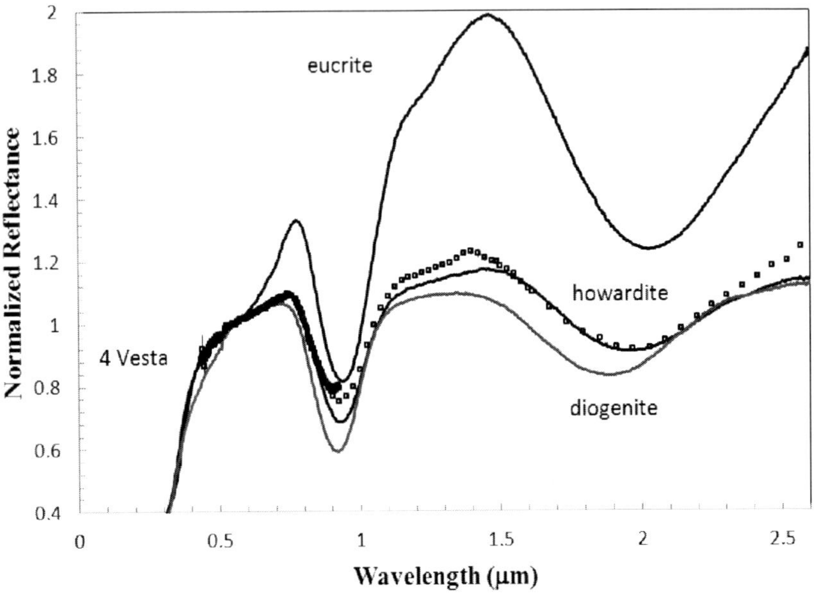

Figure 8. Reflectance spectrum of 4 Vesta (Bus and Binzel 2002a; Bell et al. 1995) versus the reflectance spectra of HEDs (Hiroi et al. 1995). Spectra are normalized to unity at 0.55 um.

This 0.7 μm feature has been found in the spectra of a number of C-type asteroids (Vilas and Gaffey 1989; Bus and Binzel 2002a). Bus and Binzel (2002b) labeled these objects as either Ch- or Cgh-types due to the presence of this feature. The difference between the Ch- and Cgh-types is that the Cgh-types have stronger UV absorptions. Burbine (1998) found that CM chondrites were spectrally similar to both main-belt asteroid 13 Egeria and 19 Fortuna (Fig. 9). Both Egeria and Fortuna are both located relatively near the meteorite-supplying 3:1 resonance and should supply a large number of fragments into the resonance (Farinella et al. 1993).

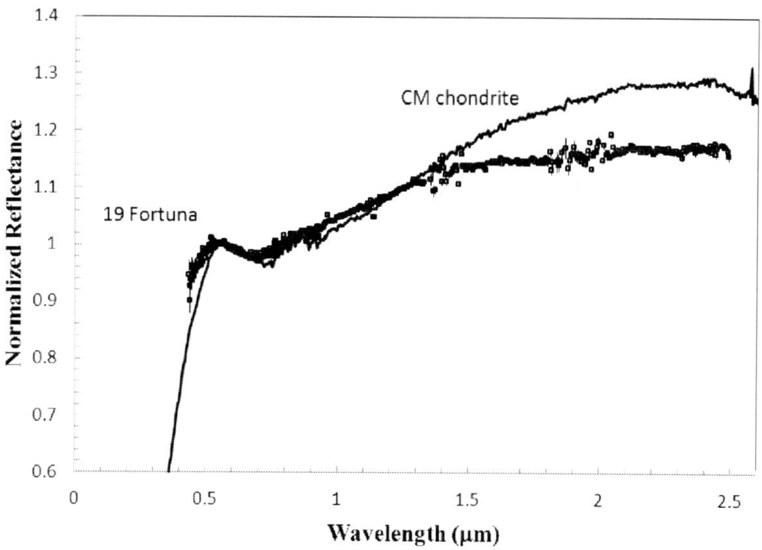

Figure 9. Reflectance spectrum of 19 Fortuna (http://smass.mit.edu/minus.html) versus the reflectance spectrum of a CM chondrite (Hiroi et al. 1993). Spectra are normalized to unity at 0.55 um.

Iron Meteorites and M-Type Asteroids

M-type asteroids have been historically linked with iron meteorites due to both types of objects having relatively flat spectra in the visible and near-infrared and moderate visual albedos (~0.10-0.30). However, radar analyses of asteroids can actually give an indication of metal content. Radar albedos are a complicated function of the near-surface bulk density (which is a function of the solid rock density and the porosity of the surface) and the mineralogy of the object. Increasing the metal content of the surface or decreasing the surface porosity will increase the radar albedo of an object. Seven M-type asteroids (16 Psyche, 129 Antigone, 216 Kleopatra, 347 Pariana, 758 Mancunia, 779 Nina, 785 Zwetana) have been found to have high radar albedos (e.g., Shepard et al. 2010) that that appear to almost certainly indicate metallic iron-rich surfaces. However, most of these asteroids have absorption features at ~0.9 and ~1.9 μm, which indicate the presence of Fe-bearing silicates. Shepard et al. (2010) believes that most of these objects are collisional composites of metallic iron and silicates, not pure metallic iron cores. Figure 10 compares the reflectance spectrum of 16 Psyche (http://smass.mit.edu/minus.html) with the reflectance spectrum of an iron meteorite (Cloutis et al. 1990).

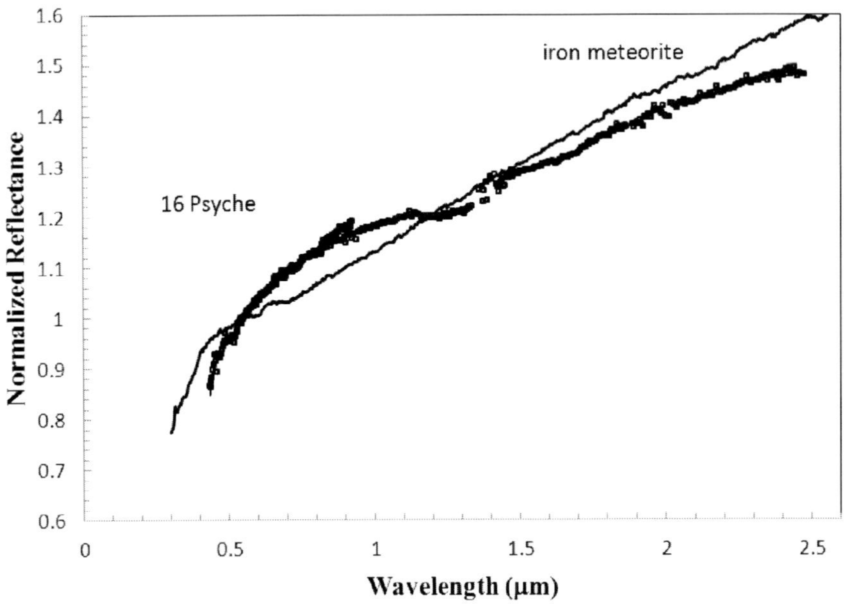

Figure 10. Reflectance spectrum of 16 Psyche (http://smass.mit.edu/minus.html) versus the reflectance spectrum of an iron meteorite (Cloutis et al. 1990). Spectra are normalized to unity at 0.55 um.

Aubrites and E-Types

E-type asteroids have flat and featureless spectra in the visible and near-infrared but have high albedos (usually above 0.30). Aubrites (e.g., Gaffey 1976; Cloutis and Gaffey 1993) also have flat and featureless spectra in the visible and near-infrared since they predominately contain enstatite (Fe-poor silicates) (Watters and Prinz 1979). E-type near-Earth asteroid 3103 Eger has been proposed to be the source body of the aubrites (Gaffey et al. 1992). Some E-types, labeled as Xe by Bus and Binzel (2002b), have an absorption band centered at ~0.49. This band has been proposed to be due to the mineral oldhamite (CaS) (Burbine et al. 20002b), a mineral found in aubrites. However, the low concentrations of oldhamite in aubrites has led Shestopalov et al. (2010) to propose that the feature is due to a titanium-bearing pyroxene that could potentially be found in very reduced enstatite material. The similarities between a reflectance spectrum of 64 Angelina (Bus and Binzel 2002a; Bell et al. 1995) and a reflectance spectrum of an aubrite (Cloutis and Gaffey 1993) are shown in Figure 11.

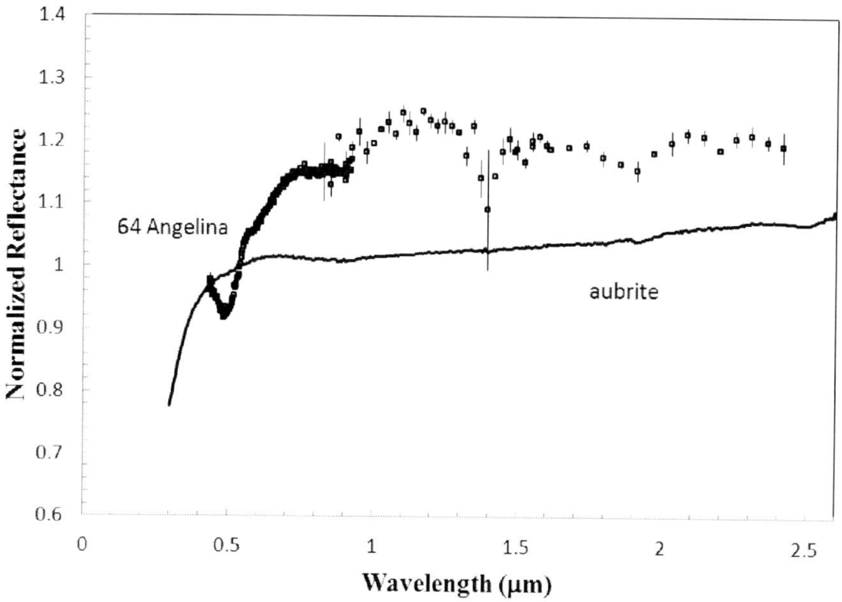

Figure 11. Reflectance spectrum of 64 Angelina (Bus and Binzel 2002a; Bell et al. 1995) versus the reflectance spectrum of an aubrite (Cloutis and Gaffey 1993). The features longwards of 1 μm in the Angelina spectrum are due to atmospheric absorptions. Spectra are normalized to unity at 0.55 um.

Ureilites and F-Types

Near-Earth asteroid 2008 TC_3 collided with the Earth's atmosphere in October of 2008 (Jenniskens et al. 2009). Before the collision, a visible reflectance spectrum was obtained and the asteroid was classified as an F-type, a subtype of the C-class asteroids with a relative flat visible spectrum with a slight bluish (reflectance decreasing with increasing wavelength) slope longwards of 0.7 μm. Luckily, fragments rained down over the Sudan and were recovered. These meteorites were found to be ureilites (Jenniskens et al. 2009) and were named the Almahata Sitta meteorites. The ureilite parent bodies were commonly assumed to be among the S-asteroids. However, the

Cloutis et al. (2010) spectroscopic study of ureilites showed that these meteorites tended to have very suppressed bands with blue spectral slopes longwards of 0.7 μm that were similar in spectral shape and structure to carbonaceous chondrites.

CONCLUSION

The search for meteorite parent bodies has dominated asteroid spectroscopic studies for the past several decades. Though uncertainties in optical, chemical, and physical properties of asteroid surfaces make it difficult to decisively link most asteroids with particular meteorite groups, several meteorite-asteroid connections have been made using telescopic observations of asteroids and laboratory analyses of meteorites in the VINIR wavelength range. One of the strongest links between meteorites and their parent asteroids is the ordinary chondrites and the S-type asteroids. The S(IV) asteroid 6 Hebe has been hypothesized to be the parent body of the H chondrites (Gaffey and Gilbert 1998), and the Flora Family has been suggested as possible source of the L chondrites (Nesvorný et al. 2002). The relationship between the S-type asteroids and the ordinary chondrites has been confirmed by the Hyabusa mission, which returned fragments of S-asteroid 25143 Itokawa in June of 2010. Laboratory analyses of these fragments were consistent with LL chondrites (Nakamura et al. 2011). Another well-supported meteorite-parent body link is that between the HED meteorites and asteroid 4 Vesta. Asteroid 4 Vesta has been linked to the HED meteorites since the first visible spectrum of Vesta (McCord et al. 1970) was obtained. The visible spectrum of Vesta was similar to the distinctive spectra of the pyroxene-rich basaltic HEDs. This relationship is currently being tested by the Dawn spacecraft, which entered into Vesta's orbit in July of 2011.

Several additional meteorite-parent body relationships have been established using VISNIR asteroid spectroscopy including one between the CM chondrites and the C-type asteroids, which is based on the presence of an absorption feature at 0.7 μm found in the spectra of numerous C-type asteroids (Vilas and Gaffey 1989; Hiroi et al. 1993). M-type asteroids have been historically linked with iron meteorites due to both types of objects having relatively flat spectra in the visible and near-infrared and moderate visual albedos (~0.10-0.30). Several M-type asteroids have been found to have high radar albedos (e.g., Shepard et al. 2010) that that appear to almost certainly indicate metallic iron-rich surfaces, though many of these asteroids have absorption features that indicate the presence of Fe-bearing silicates. The E-type asteroids and the aubrites have also been linked, as both have flat, featureless VISNIR spectra and high albedos (e.g., Gaffey 1976; Cloutis and Gaffey 1993). E-type near-Earth asteroid 3103 Eger has been proposed to be the source body of the aubrites (Gaffey et al. 1992). In 2008, a rare opportunity to test a potential meteorite-parent body link presented itself when the asteroid 2008 TC_3 impacted Earth. Recovered meteorite fragments from the 2008 TC_3 collision were found to be ureilites (Jenniskens et al. 2009). Prior to the impact of 2008 TC_3, the ureilite parent bodies were commonly assumed to be S-asteroids. However, a visible reflectance spectrum of 2008 TC_3 obtained prior to its collision with Earth suggested that it was an F-type asteroid. This experience reminds us that there is still a significant degree of uncertainty involved in establishing possible meteorite-asteroid relationships that cannot be disregarded.

ACKNOWLEDGMENTS

Part of the data utilized in this publication were obtained and made available by the MIT-UH-IRTF Joint Campaign for NEO Reconnaissance. The IRTF is operated by the University of Hawaii under Cooperative Agreement no. NCC 5-538 with the National Aeronautics and Space Administration, Office of Space Science, Planetary Astronomy Program. The MIT component of this work is supported by NASA grant 09-NEOO009-0001, and previously by the National Science Foundation under Grant No. 0506716. Special thanks to Tom Burbine for his thoughtful comments and suggestions, which greatly improved this manuscript, and for providing several of the figures included in this chapter.

Reviewed by Thomas Burbine, Mount Holyoke College, South Hadley, MA 01075.

REFERENCES

Abe, M.; Takagi, Y.; Kitazato, K.; Abe, S.; Hiroi, T.; Vilas, F.; Clark, B. E.; Abell, P. A.; Lederer, S. M.; Jarvis, K. S.; Nimura, T.; Ueda, Y.; Fujiwara, A. Near-Infrared Spectral Results of Asteroid Itokawa from the Hayabusa Spacecraft. *Science*. 2006, 312, 1334-1338.

Abell, P. A.; Vilas, F.; Jarvis, K. S.; Gaffey, M. J.; Kelley, M. S. Mineralogical composition of (25143) Itokawa 1998 SF36 from visible and near-infrared reflectance spectroscopy: Evidence for partial melting. *Meteorit. Planet. Sci.* 2007, 42, 2165-2177.

Adams, J.B. Visible and near-infrared diffuse reflectance spectra of pyroxenes as applied to remote sensing of solid objects in the solar system. *J. Geophys. Res.* 1974, 79, 4829-4836.

Adams, J.B. Interpretation of visible and near-infrared diffuse reflectance spectra of pyroxene and other rock-forming minerals. In *Infrared and Raman Spectroscopy of Lunar and Terrestrial Minerals*; Karr, C.; Academic Press: New York: NY, 1975; pp. 91-116.

Anders, E.; Grevesse, N. Abundances of the elements: meteoritic and solar. *Geochim Cosomchim Acta*. 1989, 53, 197-214.

Barucci, M.A.; Capria, M.T.; Coradini, A.; Fulchignoni, M. Classification of asteroids using G-mode analysis. *Icarus*. 1987, 72, 304-324.

Bell, J.F. A Probable Asteroidal Parent Body for the CV or CO Chondrites. *Meteoritics*. 1988, 23, 256.

Bell, J.F.; Owensby, P.D.; Hawke, B.R.; Brown, R.H.; Cruikshank, D.; Hartmann, W.K. 52-color Asteroid Survey. EAR-A-RDR-3-52COLOR-V2.1. *NASA Planetary Data System,* 1995.

Benedix, G.K; McCoy, T.J.; Keil, K.; Bogard, D.D.; Garrison, D.H. A petrologic and isotopic study ofwinonaites: Evidence for early partial melting and crystallization. *Geochim Cosomchim Acta*.1998, 62, 2535-2553.

Binzel, R.P.; Xu, S. Chips off of asteroid 4 Vesta - Evidence for the parent body of basaltic achondrite meteorites. *Science*. 1993, 260, 186-191.

Binzel, R. P.; Rivkin, A. S.; Bus, S. J.; Sunshine, J. M.; Burbine, T. H. MUSES-C target asteroid 1998 SF36: A reddened ordinary chondrite. *Meteorit. Planet. Sci.* 2001, 36, 1167-1172.

Binzel, R. P.; Morbidelli, A.; Merouane, S.; DeMeo, F. E.; Birlan, M.; Vernazza, P.; Thomas, C. A.; Rivkin, A.S.; Bus, S. J.; Tokunaga, A. T. (2010) Earth encounters as the origin of fresh surfaces on near-Earth asteroids. *Nature*. 2010, 463, 331-334.

Brown, P.G.; Hildebrand, A.R.; Zolensky, M.E.; Grady, M.; Clayton, R.N.; Mayeda, T.K.; Tagliaferri, E.; Spalding, R.; MacRae, N.D.; Hoffman, E.L.; Mittlefehldt, D.W.; Wacker, J.F.; Bird, J.A.; Campbell, M.D.; Carpenter, R.; Gingerich, H.; Glatiotis, M.; Greiner, E.; Mazur, M.J.; McCausland, P.J.A.; Plotkin, H.; Rubak M,T. The Fall, Recovery, Orbit, and Composition of the Tagish Lake Meteorite: A New Type of Carbonaceous Chondrite. *Science*. 2000, 290, 320-325.

Burbine, T.H. Could G-class asteroids be the parent bodies of the CM chondrites?. *Meteorit. Planet. Sci.* 1998, 33, 253-258.

Burbine, T. H.; Buchanan, P. C.; Binzel, R. P.; Bus, S. J.; Hiroi, T.; Hinrichs, J. L.; Meibom, A.; McCoy, T. J. Vesta, Vestoids, and the HEDs: Relationships and the origin of spectral differences. *Meteorit. Planet. Sci.* 2001a, 36, 761-781.

Burbine, T. H.; McCoy, T. J.; Nittler, L. R.; Bell, J. F. III. Could 433 Eros have a primitive achondritic composition? (abstract). *Lunar and Planetary Science*. 2001b, 32, 1860.

Burbine, T.H.; McCoy T.J.; Meibom, A.; Gladman, B.; Keil, K. Meteoritic parent bodies: Their number and identification. In *Asteroids III;* Bottke, W.F. Jr.; ellino, A.; Paolicchi, P.; Binzel, R.P; The University of Arizona Press: Tucson, AZ, 2002a; pp. 653-667.

Burbine T.H.; McCoy T.J.; Nittler L.R.; Benedix G.K.; Cloutis E.A.; Dickinson T.L. Spectra of extremely reduced assemblages: Implications for Mercury. *Meteorit. Planet. Sci.* 2002b, 37, 1233-1244

Burbine, T. H.; McCoy, T. J.; Jarosewich, E.; Sunshine, J. M. Deriving asteroid mineralogies from reflectance spectra: Implications for the Muses-C target asteroid. *Antarct. Meteor. Res.* 2003, 16, 185-195.

Burbine, T.H.; Buchanan, P.C.; Binzel, R.P. Deriving formulas from HED spectra for determining the pyroxene mineralogy of Vesta and vestoids (abstract #2117). *38th Lunar and Planetary Science Conf.* 2007. (CD-ROM)

Burns, R.G. Crystal field spectra and evidence of cation ordering in olivine minerals. *Am. Mineral.* 1970, 55, 1608-1632.

Bus, S.J. *Compositional structure in the asteroid belt: Results of a spectroscopic survey.* PHD Thesis. 1999, Massachusettes Institute of Technology, Boston, MA.

Bus, S. J.; Binzel, R. P. Phase II of the Small Main-belt Asteroid Spectroscopic Survey: The observations. *Icarus*. 2002a, 158,106-145.

Bus, S.J.; Binzel, R.P. Phase II of the Small Main-Belt Asteroid Spectroscopic Survey: A feature-based taxonomy. *Icarus*. 2002b, 158, 146-177.

Chapman, C. R.; Morrison, D.; Zellner, B. Surface properties of asteroids - A synthesis of polarimetry, radiometry, and spectrophotometry. *Icarus*. 1975, 25, 104-130.

Chizmadia, L.J.; Rubin, A.E.; Wasson, J.T. Mineralogy and petrology of amoeboid olivine inclusions in CO3 chondrites: Relationship to parent-body aqueous alteration. *Geochim Cosomchim Acta*. 2002, 37, 1781-1796.

Clark, B.E; Lucey, P.; Helfenstein, P. The Nature of Space Weathering Effects on Asteroid 433 Eros. *Solar System Remote Sensing*. 2002, 13.

Clark, S.P. Jr. Absorption spectra of some silicates in the visible and near-infrared. *Am. Mineral.* 1957, 42, 732-742.

Clayton, R.N. Oxygen isotopes in meteorites. *Ann. Rev. of Earth and Planet. Sci.* 1993, 21,115-149.

Clayton, R.N; Mayeda, T.K. Oxygen isotope studies of achondrites. *Geochim Cosomchim Acta.* 1996, 60, 1999-2018.

Cloutis, R.N. *Interpretive techniques for reflectance spectra of mafic silicates.* MS Thesis. 1985, University of Hawaii, Honolulu, HI.

Cloutis, E. A.; Gaffey, M. J.; Jackowski, T. L.; Reed, K. L. Calibrations of phase abundance, composition, and particle size distribution of olivine-orthopyroxene mixtures from reflectance spectra. *J. Geophys. Res.* 1986, 91, 11641-11653.

Cloutis E. A.; Gaffey M. J.; Smith D. G. W.; Lambert R. St. J. Reflectance spectra of “featureless” materials and the surface mineralogies of M- and E-class asteroids. *J. Geophys. Res.* 1990, 95,281–293.

Cloutis, E.A.; Gaffey, M.J. Pyroxene spectroscopy revisited: Spectral-compositional correlations and relationships to geothermometry. *J. Geophys. Res.* 1991, 91, 11641-11653.

Cloutis E.A.; Gaffey M. J. Accessory phases in aubrites: Spectral properties and implications for asteriod 44 Nysa. *Earth, Moon, and Planets.* 1993, 63, 227-243

Cloutis, E. A.; Hudon, P.; Romanek, C. S.; Bishop, J. L.; Reddy, V.; Gaffey, M. J.; Hardersen, P. S. Spectral reflectance properties of ureilites. *Meteorit. Planet. Sci.* 2010, 45, 1668-1694.

Consolmagno, G.J.; Drake, M.J. Composition and evolution of the eucrite parent body - Evidence from rare earth elements. *Geochim Cosmochim Acta*. 1977, 41, 1271-1282.

Cruikshank, D.P; Hartmann, W.K. The meteorite-asteroid connection - Two olivine-rich asteroids. *Science*. 1984, 223, 281-283.

Delaney, J.S.; Prinz, M.; Takeda, H. The polymict eucrites. *J. Geophys. Res.* 1984, 89, C251-C288.

DeMeo, F.E.; Binzel, R.P.; Slivan, S.M.; Bus, S. J. An extension of the Bus asteroid taxonomy into the near-infrared. *Icarus*. 2009, 202, 160-180.

Drake. M.J. The eucrite/Vesta story. *Meteorit. Planet. Sci.* 2001, 4, 501-513.

Dunn, T. L.; Cressey, G.; McSween, H. Y. Jr.; McCoy, T. J.; Bland, P. A. Analysis of ordinary chondrites using powder X-ray diffraction: 1. Modal mineral abundances. *Meteorit. Planet. Sci.* 2010a, 45, 123-134.

Dunn, T.L.; McSween, Jr., H.Y.; McCoy, T.J.; Cressey, G. Analysis of ordinary chondrites using power X-ray diffraction: 2. Applications to ordinary chondrite parent-body processes. *Meteorit. Planet. Sci.* 2010b, 45, 135-156.

Dunn, T. L.; McCoy, T. J.; Sunshine J. M.; McSween H. Y. Jr. A coordinated spectral, mineralogical, and compositional study of ordinary chondrites. 2010c. *Icarus*, 208, 789-797.

Farinella, P.; Gonczi, R.; Froeschle, Ch.; Froeschle, C. The injection of asteroid fragments into resonances. *Icarus*. 1993, 101, 174-187.

Fredriksson, K.; Keil, K. The light-dark structure in the Pantar and Kapoeta stone meteorites. *Geochim. Cosomchim. Acta.* 1963, 27, 717-739.

Foley, C. N.; Nittler, L. R.; McCoy, T. J.; Lim, L. F.; Brown, M. R. M.; Starr, R. D.; Trombka, J. I. Minor element evidence that Asteroid 433 Eros is a space-weathered ordinary chondrite parent body. *Icarus*. 2006, 184, 338-343.

Gaffey, M.J. Spectral reflectance characteristics of the meteorite classes. *J.Geophys. Res*. 1976, 81, 905-920.

Gaffey, M.J.; McCord, T.B. Asteroid surface materials - Mineralogical characterizations from reflectance spectra. *Space Sci. Rev*. 1978, 21, 555-628.

Gaffey, M.J.; Reed, K.L.; Kelley, M.S. Relationship of E-type Apollo asteroid 3103 (1982 BB) to the enstatite achondrite meteorites and the Hungaria asteroids. *Icarus*. 1992, 100, 95-109.

Gaffey, M. J.; Bell, J. F.; Brown, R. H.; Burbine, T. H.; Piatek, J.; Reed, K. L.; Chaky, D. A. Mineralogic variations within the S-type asteroid class. *Icarus*. 1993, 106, 573-602.

Gaffey, M.J.; Gilbert, S.L. Asteroid 6 Hebe: The probable parent body of the H-type ordinary chondrites and the IIE iron meteorites. *Meteorit. Planet. Sci.* 1998, 33, 1281-1295.

Gaffey, M. J.; Cloutis, E. A.; Kelley, M. S.; Reed, K. L. Mineralogy of asteroids. In *Asteroids III*; Bottke, W.; Cellino, A.; Paolicchi, P.; Binzel, R. P.; University of Arizona Press, Tucson, AZ, 2002, pp. 183-204.

Gaffey, M. J. Space weathering and the interpretation of asteroid reflectance spectra. *Icarus*. 2010, 209, 564-574.

Goodrich, C.A.; Jones, R.H.; Berkley, J.L. Origin and evolution of the ureilite parent magmas: Multistage igneous activity on a large parent body. *Geochim. Cosomchim. Acta.* 1987, 51, 2255-1173.

Goodrich, C.A.; Fioretti, A.M.; Tribaudino, M.; Molin, G. Primary trapped melt inclusions in olivine in the olivine-augite-orthopyroxene ureilite Hughes 009. *Geochim. Cosomchim. Acta.* 2001, 65, 621- 652.

Goodrich, C.A.; Krot, A.E.; Scott E.R.D.; Taylor, G.J.; Fioretti, A.M.; Keil, K. Formation and evolution of the ureilite parent-body and its offspring (abstract). In *Lunar and Planetary Sciences XXXIII*, Abstract #1379. 2002, Lunar and Planetary Institute, Houston. (CD-ROM).

Gradie, J. C.; Veverka, J.; Buratti, B. J. The Effects of Photometric Geometry on Spectral Reflectance. *Lunar Planet. Sci. Conf.* 1980, 11, 357-359.

Haack, H.; McCoy, T.J. Iron and Stony-iron meteorites. In *Treatise on Geochemistry, Vol 1. Meteorites,* Comets, and Planets; Davis, A.M.; Elsevier: Oxford, England, 2004, pp. 325-345.

Hapke, B. Space weathering from Mercury to the asteroid belt. *J. Geophys. Res.*2001, 106, 10039-10074.

Hardersen, Paul S.; Gaffey, Michael J.; Abell, Paul A. Mineralogy of Asteroid 1459 Magnya and implications for its origin. *Icarus*. 2004, 167, 170-177.

Hinrichs, J. L.; Lucey, P. G.; Robinson, M. S.; Meibom, A.; Krot, A. N. Implications of temperature-dependent near-IR spectral properties of common minerals and meteorites for the remote sensing of asteroids. *Geophys. Res. Letts.* 1999, 26, 1661-1664.

Hiroi, T. and Takeda H. Reflectance spectroscopy and mineralogy of primitive achondrites-lodranites. Proc. NIPR Symp. *Antarct. Meteorites.* 1991, 4, 163-177.

Hiroi,T.; Pieters, C.M.; Zolensky, M.E. Comparison of reflectance spectra of C asteroids and unique C chondrites Y86720, Y82162, and B7904. *In Lunar and Planetary Inst.,* Twenty-Fourth Lunar and Planetary Science Conference. 1993, 659-660.

Hiroi T.; Pieters C.M..; Zolensky M.E.; Lipschutz M.E. Evidence of thermal metamorphism on the C, G, B, and F asteroids. *Science*. 1993, 261, 1016-1018.

Hiroi, T.; Pieters, C.M.; Hiroshi, T. Grain size of the surface regolith of asteroid 4 Vesta estimated from its reflectance spectrum in comparison with HED meteorites. *Meteoritics*. 1994, 29, 394-396.

Hiroi, T.; Binzel, R.P.; Sunshine, J.M.; Pieters, C.M.; Takeda, H. Grain sizes and mineral compositions of surface regoliths of Vesta-like asteroids. *Icarus*. 1995, 115, 374-386.

Hiroi, T.; Zolensky, M.E.; Pieters, C.M.; Lipschutz, M.E. Thermal metamorphism of the C, G, B, and F asteroids seen from the 0.7 micron, 3 micron and UV absorption strengths in comparison with carbonaceous chondrites. *Meteorit. Planet. Sci.* 1996, 31, 321-327.

Hiroi, T.; Zolensky, M.E.; Pieters, C.M. The Tagish Lake Meteorite: A Possible Sample from a D-Type Asteroid. *Science.* 2001, 293, 2234-2236.

Howell, E.S.; Merenyi, E.; Lebofsky, L.A. Classification of asteroid spectra using a neural network. *J.Geophys. Res.* 1994, 99, 10847-10865.

Hutchison, R. *Meteorites: A Petrologic, Chemical and Isotopic Synthesis;* Cambridge University Press; New York, NY, 2004.

Jarosewich, E. Chemical analyses of meteorites: A compilation of stony and iron meteorite analyses. *Meteoritics.* 1990, 25, 323-337.

Jenniskens, P.; Shaddad, M. H.; Numan, D.; Elsir, S.; Kudoda, A. M.; Zolensky, M. E.; Le, L.; Robinson, G.A.; Friedrich, J. M.; Rumble, D.; Steele, A.; Chesley, S. R.; Fitzsimmons, A.; Duddy, S.; Hsieh, H. H.; Ramsay, G.; Brown, P. G.; Edwards, W. N.; Tagliaferri, E.; Boslough, M. B.; Spalding, R. E.; Dantowitz, R.; Kozubal, M.; Pravec, P.; Borovicka, J.; Charvat, Z.; Vaubaillon, J.; Kuiper, J.; Albers, J.; Bishop, J. L.; Mancinelli, R. L.; Sandford, S. A.; Milam, S. N.; Nuevo, M.; Worden, S. P. The impact and recovery of asteroid 2008 TC3. *Nature.* 2009, 458, 485-488.

Keil, K. Enstatite meteorites and their parent bodies. *Meteoritics.* 1989, 24, 195-208.

King, T.V.; Ridley, W.I. Relation of the spectroscopic reflectance of olivine to mineral chemistry and some remote sensing applications. *J. Geophys. Res.* 1987, 92, 11457-11469.

Larson, H. P.; Fink, U. Infrared spectral observations of asteroid 4 Vesta. *Icarus.* 1975, 26, 420-427.

Lazzaro, D.; Michtchenko, T.; Gumarães, T. A.; Angeli, C. A. Spectroscopic and Dynamic Analysis of the Region around the Basaltic Asteroid 1459 Magnya. *Bull. Amer. Astronom. Soc.* 2000, 32, 100.

Lovering, J.F.; Nichiporuk, W.; Chodos, A.; Brown, H. The distribution of gallium, germanium, cobalt, chromium, and copper in iron and stony-iron meteorites in relation to Ni content and structure. *Geochim. Cosomchim. Acta.* 1957, 11, 263-278.

Lim, L. F.; Nittler, L. R. Elemental composition of 433 Eros: New calibration of the NEAR-Shoemaker XRS data. *Icarus.* 2009, 200, 129-146.

Lucey, P.G.; Keil, K.; Whitely, R. The influence of temperature on the spectra of the A-asteroids and implications for their silicate chemistry. *J. Geophys. Res.* 1998, 103, 5865.

McCord, Thomas B.; Adams, John B.; Johnson, Torrence V. Asteroid Vesta: Spectral Reflectivity and Compositional Implications. Science. 1970, 168, 1445-1447.

McCoy, T.J.; Keil, K.; Clayton, R.N.; Mayeda, T.K.; Bogard D.D.; Garrison D.H.; Huss, G.R.; Hutcheon I.D.; Wieler R. A petrologic, chemical, and isotopic study of Monument Draw and comparison with other acapulcoites: Evidence for formation by incipient partial melting. *Geochim. Cosomchim. Acta.* 1996, 60, 2681-2708.

McCoy, T.J.; Keil, K.; Clayton, R.N.; Mayeda, T.K.; Bogard D.D.; Garrison D.H;Wieler R. petrologic and isotopic study of lodranites: Evidence for early formation as partial melt residues from heterogeneous precursors. *Geochim. Cosomchim. Acta.* 1997, 61, 623-637.

McCoy, T. J.; Nittler, L. R.; Burbine, T. H.; Trombka, J. I.; Clark, P. E.; Murphy, M.E. Anatomy of a partially differentiated asteroid: A "NEAR"-sighted view of acapulcoites and lodranites. *Icarus.* 2000, 148, 29-36.

McCoy, T.J.; Burbine, T.H.; McFadden, L.A.; Starr, R.D.; Gaffey, M.J.; Nittler, L.R.; Evans, L.G.; Izenberg, N.; Lucey, P.; Trombka, J.I.; Bell, J.F., III; Clark, B E.; Clark, P E.; Squyres, S.W.; Chapman, C.R.; Boynton, W.V.; Veverka, J. The composition of 433 Eros: A mineralogical-chemical synthesis. *Meteorit. Planet. Sci.* 2001, 36,1661-1672.

McFadden, L. A.; Pieters, C.; McCord, T. B. Vesta - The first pyroxene band from new spectroscopic measurements. *Icarus.* 1977, 31, 439-446.

McSween, H.Y. Jr. Petrologic variations among carbonaceous chondrites of the Vigarano type. *Geochim Cosomchim Acta.* 1977, 57, 1105-1114.

McSween, H.Y. Jr.; Sears, D.W.G.; Dodd, R.T. In *Meteorites and the Early Solar System;* Kerridge, J.F. and Matthews, M.S.; University of Arizona Press: Tucson, AZ, 1988; pp. 102-113.

McSween, H.Y. Jr. What we have learned about Mars from SNC meteorites. *Meteoritics.* 2004. 29, 757-779.

Mittlefehldt, D.W.; Lindstrom, M.M. Geochemistry and genesis of the angrites. *Geochim. Cosomchim. Acta.* 1990, 54, 3209-3218

Mittlefehldt , D.W.; Lindstrom, M.M.; Bogard, D.D.; Garrison, D.H.; Field, S.W. Acapulco- and Lodran-like achondrites: Petrology, geochemistry, chronology and origin. *Geochim. Cosomchim. Acta.* 1996, 60, 867-882.

Mittlefehldt,D.W.; McCoy, T.J.; Goodrich, C.A.; Kracher, A. Non-chondritic meteorites from asteroidal bodies. In *Planetary Materials*; Papike, J.J.; Reviews in Mineralogy, 36th ed.; Mineralogical Society of America: Washington, DC, 1998: pp 4-1 – 4-195.

Mittlefehldt, D.W. Achondrites. In Treatise on Geochemistry, Vol 1. *Meteorites, Comets, and Planets*; Davis, A.M.; Elsevier: Oxford, England, 2004; pp. 291-324.

Moroz, L.; Schade, U.; Wäsch, R. Reflectance spectra of olivine–orthopyroxene-bearing assemblages at decreased temperatures: Implications for remote sensing of asteroids. *Icarus*. 2000, 147, 79-93.

Muinonen, K.; Piironen, J.; Shkuratov, Yu. G.; Ovcharenko, A.; Clark, B. E. Asteroid Photometric and Polarimetric Phase Effects. In *Asteroids III*, W. F. Bottke Jr., A. Cellino, P. Paolicchi, and R. P. Binzel, 2002, University of Arizona Press, Tucson, p.123-138

Nagahara, H. Yamato-8002: Partial melting residue on the "unique" chondrite parent body. Proc. NIPR Sympo. *Antarct. Meteorites*. 1992, 5, 191-223.

Nakamura, T.; Noguchi, T.; Tanaka, M.; Zolensky, M.E.; Kimura, M.; Nakato, A.; Ogami, T.; Ishida, H.; Tsuchiyama, A.; Yada, T.; Shirai, K.; Okazaki, R.; Fujimura, A.; Ishibashi, Y.; Abe, M.; Okada, T.; Ueno, M.; Mukai, T. Mineralogy and Major Element Abundance of the Dust Particles Recovered from Muses-C Regio on the Asteroid Itokawa. *42nd Lunar and Planetary Science Conference,* 2011, LPI Contribution No. 1608, p.1766

Nehru, C.E.; Prinz, M.; Weisberg, M.K.; Ebihara, M.E.; Clayton, R.E. Mayeda T.K. brachinites: A new primitive achondrite group. *Meteoritics*. 1996, 27, 267.

Nesvornỳ, D.; Morbidelli, A.; Vokrouhlicky, D.; Bottke, W.F.; Broz, M. The flora family: A case of the dynamically dispersed collisional swarm? *Icarus*. 2002, 157, 155-172.

Prinz, M.; Keil, K.; Hlava, P.F.; Berkley, J.L.; Gomes, C.B.; Curvello, W.S. Studies of Brazilian meteorites, III. Origin and history of the Angra dos Reis achondrite. *Earth Planet. Sci. Lett*. 1977, 35, 317-330.

Rayner, J. T.; Toomey, D. W.; Onaka, P. M.; Denault, A. J.; Stahlberger, W. E.; Vacca, W. D.; Cushing, M.C.; Wang, S. SpeX: A medium-resolution 0.8-5.5 micron spectrograph and imager for the NASA Infrared Telescope Facility. *Pub. Astronom. Soc. Pacific*. 2003, 115, 362-382.

Roig, F.; Nesvorný, D.; Gil-Hutton, R.; Lazzaro, D. V-type asteroids in the middle main belt. *Icarus*. 2008, 194, 125-136.

Roush, T. L.; Singer, R. B. Possible temperature variations effects on the interpretation of spatially resolved reflectance observations of asteroid surfaces. *Icarus*. 1987, 69, 571-574.

Sasaki, S.; Nakamura, K.; Hamabe, Y.; Kurahashi, E.; Hiroi, T. Production of iron nanoparticles by laser irradiation in a simulation of lunar-like space weathering. *Nature*. 2001, 410, 555-557.

Sears, D.W.G.; Grossman, J.N.; Melcher, C.L.; Ross, L.M.; Mills, A.A. Measuring metamorphic history of unequilibrated ordinary chondrites. *Nature*. 1980, 287, 791-795.

Sears, D.W.G.; Hasan, F.A.; Batchelor, J.D.; Lu, J. Chemical and physical studies of type 3 chondrites: XI, Metamorphism, paring, and brecciation of ordinary chondrites. *Lunar Planet. Sci. Conf.* 1991, 22, 493-512.

Shepard, M.K.; Clark, B.E; Ockert-Bell, M.; Nolan, M.C.; Howell, E.S.; Magri, C.; Giorgini, J.D.; Benner, L. A. M.; Ostro, S.J.; Harris, A.W.; Warner, B.D.; Stephens, R.D.; Mueller, M. A radar survey of M- and X-class asteroids II. Summary and synthesis. *Icarus*. 2010, 1, 221-237.

Shestopalov, D. I.; Golubeva, L. F.; McFadden, L. A.; Fornasier, S.; Taran, M. N. Titanium-bearing pyroxenes of some E asteroids: Coexisting of igneous and hydrated rocks. *Planet. Space. Sci*. 2010, 58, 1400-1403.

Singer, R. B; Roush, T. L. Effects of temperature on remotely sensed mineral absorption features. *J. Geophys. Res*. 1985, 90, 12434-12444.

Strazzulla, G.; Dotto, E.; Binzel, R.; Brunetto, R.; Barucci, M. A.; Blanco, A.; Orofino, V. Spectral alteration of the meteorite Epinal (H5) induced by heavy ion irradiation: a simulation of space weathering effects on near-Earth asteroids. *Icarus* 2005, 174, 31-35.

Sunshine, J.M.; Pieters, C.M.; Determining the composition of olivine from reflectance spectroscopy. J. Geophys. Res. 1998, 103, 13675-13688.

Tedesco, E.F.; Williams, J.G.; Matson, D.L.; Veeder, G.L; Gradie, J.C.; Lebofsky, L.A. A three-parameter asteroid taxonomy. *Astron. J*. 1989, 97, 580-606.

Tedesco, E.F.; Cellino, A.; Zappalá, V. The Statistical Asteroid Model. I. The Main-Belt Population for Diameters Greater than 1 Kilometer. *Astron. J.*2005, 129, 2869-2886.

Tholen, D.J. *Asteroid taxonomy from cluster analysis of photometry*. PhD Thesis. 1984.University of Arizona, Tucson, AZ.

Thomas, C. A.; Binzel, R. P. Identifying meteorite source regions through near-Earth spectroscopy. *Icarus*. 2010, 205, 419-429.

Trombka, J. I.; Squyres, S. W.; Brückner, J.; Boynton, W. V.; Reedy, R. C.; McCoy, T. J.; Gorenstein, P.; Evans, L. G.; Arnold, J. R.; Starr, R. D.;

Nittler, L. R.; Murphy, M. E.; Mikheeva, I.; McNutt, R. L Jr.; McClanahan, T. P.; McCartney, E.; Goldsten, J. O.; Gold, R. E.; Floyd, S. R.; Clark, P. E.; Burbine, T. H.; Bhangoo, J. S.; Bailey, S. H.; Petaev, M. The elemental composition of asteroid 433 Eros: Results of the NEAR Shoemaker X-Ray spectrometer. *Science*. 2000, 289, 2101-2105.

Van Schmus, W.R.; Wood, J.A. A chemical-petrologic classification for the chondritic meteorites. *Geochim. Cosmochim. Acta*. 1967, 31, 747-765.

Vernazza, P.; Binzel, R. P.; Rossi, A.; Fulchignoni, M.; Birlan, M. Solar wind as the origin of rapid reddening of asteroid surfaces. *Nature*. 2009, 458, 993-995.

Vilas, F.; Gaffey, M. J. Weak Fe2+ -Fe3+ Charge Transfer Absorption Features Seen in CM2 Carbonaceous Chondrites and Narrowband Reflectance Spectra of Primitive Asteroids. *Lunar Planet Sci. Conf.* 1989, 20, 1156.

Warren, P.H.; Kallemeyn, G.W. Allan Hills 84025: The second brachinite, far more differentiated than Brachina, and an ultramafic achondritic clast from L chondrite Yamato 75097. *Lunar. Planet. Sci. Conf.* 1989, 20, 475-486.

Warren, P. H.; Kallemeyn, G. W. Ureilites: The Graphite f(O2) Buffer, Explosive Volcanism, and the Gross Dissipation of Basalt from the Parent Asteroid(s). *Lunar Planet. Sci. Conf.* 1992, 23, 1497.

Warren, P.H. Lunar and martian meteorite delivery services. *Icarus*. 1994, 111, 338-363.

Watters, T.R.; Prinz, M. Aubrites: Their origin and relationship to enstatite chondrites. *Lunar. Planet. Sci. Conf.* 1979, 10, 1073-1093.

Weisberg, M.K.; McCoy, T.J.; Krot, A.N. Systematics and Evaluation of Meteorite Classification. In *Meteorites and the Early Solar System II:* Lauretta D.S and McSween H.Y. Jr.; University of Arizona Press: Tucson, AZ. 2006 pp. 19-52.

Xu, S.; Binzel ,R. P.; Burbine, T. H.; Bus, S. J. Small main-belt asteroid spectroscopic survey: Initial results. *Icarus*. 1995, 115, 1-35.

In: Meteorites and Asteroids
Editors: A. Dementieva et al.
ISBN: 978-1-61942-115-8

Chapter 2

EXTRATERRESTRIAL MATERIAL AND CHANGES OF THE EARTH'S CLIMATE

M. Ogurtsov
A.F. Ioffe Physico-Technical Institute of Russian Academy of Sciences, St.-Petersburg, Russia
Central Astronomic Observatory at Pulkovo, Russia

ABSTRACT

The process of extraterrestrial material income into the Earth's atmosphere is considered. It is shown that particles of meteoric smoke may serve as condensation nuclei in the troposphere and stratosphere. It is shown that charged particles of meteoric origin might appreciably change mesospheric, and, as a result, the total atmospheric resistance, which, in turn, affect the global current circuit. Changes in the global electric circuit may influence processes of aerosol and cloud formation. The possible climatic effect of interstellar dust in the Earth's atmosphere is found negligible. Possible connection between terrestrial climatic variation with a period 2.0-2.5 kyr (the Hallstatt cycle) and nodal precession of the core of Taurid meteoroid stream is analyzed. It is confirmed that extremes of the Hallstatt cycle coincide with the data of probable intersections of the Earth's orbit by the core of Taurid complex. Since during these epochs intensity of the Earth's bombardment by cometary debris increases sharply, the possible results of collision of the Earth with asteroid, which has size of 250 m and velocity of 20 km×s^{-1} is estimated. It is shown that climatic consequences of such impact might be substantial and therefore a chain of such events may contribute to 2.0-2.5 kyr climatic variation appreciably. The obtained results testify that the extraterrestrial factors actually might contribute to climatic variability over different time scales.

INTRODUCTION

A number of solid cosmic bodies – micrometeoroids (size 25-1000 microns) and meteoroids (size from 1000 microns to several meters) – continuously (at least 1 impact per year) enter the atmosphere of the Earth. The bodies of larger size (tens of meters or more) also collide with the Earth, despite less frequently. The basic source of these objects is the products from the destruction of asteroids and comet cores. The speeds with which the cosmic bodies enter the atmosphere are 11-73 km/s (Birks et al., 2007). The masses of meteoric bodies range from 10^{-9} to 10^{7} g. The majority (93 %) of meteorites falling to the ground are made of stone and the rest of them are mainly iron and iron-stone meteorites (Tirsky, 2000). Stone meteorites consist basically from olivine ($MgFeSiO_4$) and pyroxene ($Mg_{0.5}Fe_{0.5}SiO_3$) and have density ρ=3.0-3.5 g×cm^{-3}. The density of the iron meteorites is about 7.6 g×cm^{-3}. Particles of interplanetary and interstellar dust with sizes from 0.1 micron to 25-40 microns also enter the terrestrial atmosphere. As a result the Earth's atmosphere undergoes both the influence of persistent flux of extraterrestrial substance and the sporadic impacts connected with intrusions of cosmic bodies of considerable (more than tens of meters) sizes. Possible environmental and climatic consequences of the separate impacts of large asteroids have been analyzed in many works – see e.g. Bailer-Jones (2009) and references therein. It has been demonstrated that these events likely have a significant effect on both global climate and the biosphere. Moreover, the falling of giant (more than 1 km) objects could result in dramatic episodes such as mass extinctions (Alvarez et al. (1980); Pope et al. (1994); Toon et al. (1997)). However such occasions occur very seldom (< 1 per 10^4 yrs) and thus the impacts of mid-size celestial bodies (size hundreds of meters) might have a more significant influence on the Earth's environment.

Recently Ermakov et al. (2006) and Kasatkina et al. (2007) reported that the continuous flux of cosmic dust might also affect the Earth's climate. It has been assumed that dust and aerosol particles of extraterrestrial origin entering the atmosphere may serve as condensation nuclei (CN) and, thus, influence cloud cover (Ermakov, et al., 2006). Consequently, the flux of cosmic substance falling on the Earth's surface could operate as an appreciable climatic factor, which is worth some qualitative estimation.

THE INFLUENCE OF CONTINUOUS FLUX OF EXTRATERRESTRIAL SUBSTANCE ON ATMOSPHERIC PROCESSES

Available estimations of a total flux of extraterrestrial matter coming to the atmosphere of the Earth are listed in Tab. 1.

Table 1. Estimations of extraterrestrial accretion

The authors	Way of estimation	Extraterrestrial flux (Kiloton/year)
Kane and Gardner, 1993	Lidar observation	2.0±0.6
Taylor et al., 1998	Analysis of cosmic spherules in South Pole bottom sediments.	2.7±1.4 (r = 50-700 microns)
Maurett et al., 1987	Analysis of the dust particles in the Greenland ice	5 (r>50 microns)
Rasmussen et al., 1995	Analysis of the iridium concentration in the Greenland ice	10±2
Dohnanyi, 1972	Model calculations	20
Love and Brownlee, 1993	Analysis of microcraters on the surface of the satellite target	40±20
Esser and Turekian, 1988	Analysis of the osmium concentration in oceanic clay	49–56
Kyte and Wasson, 1986	Analysis of the iridium concentration in oceanic clay	60–120
Ceplecha, 1996	Model calculations	150

The evident disagreement in the estimates could be a result of considerable temporal and spatial unhomogeneity of the extraterrestrial fluxes.

At the heights of 80-130 km meteoric bodies reaching the Earth's atmosphere ablate. The process of ablation includes destruction, evaporation and dispersion of the cosmic material. Large meteors evaporate completely in the atmosphere. Micrometeors not always heat up to a boiling point (approximately 2000 ^{0}C) and thus can reach the surface of the Earth. According to the calculations of Hunten et al., (1980) bodies weighing less than 10^{-6} g (the size <50 microns) practically do not lose weight due to evaporation. It has been assumed (Rosinski and Snow, 1961; Hunten et al.,

1980) that the evaporated meteoric substance in the top of the atmosphere re-condenses forming particles of nanometer size – and thus creating meteoric dust or a smoke. Particles of the sizes 2-5 nanometers (0.002-0.005 microns) have indeed been found experimentally at heights of 70-90 km - see (Rapp et al., 2007) and references therein. Estimations of the weight of substance coming into the atmosphere in the form of meteoric vapor are rather different. In their work Kane and Gardner, (1993) estimated this weight as 2.0 kt/year, while in the work of Lal and Jull (2002) as 30 kt/year. The distribution of particle concentration of meteoric dust in the atmosphere is determined in general by model calculations (Fig. 1). The concentration of particles of meteoric dust n (z) (sm^{-3}) below 30 km is calculated as $n(z) = n_0(z) \times \rho(z)$, where $\rho(z)$ is air density, and the mixing ratio $n_0(z)$ (mg^{-1}) was taken from fig. 9a of the work by Hunten et al. (1980) and above 30 km $n(z)$ it was taken from Fig2a by Megner et al. (2006). The modal radius of particles of meteoric dust in the mesosphere is 2-5 nanometers (Rapp et al., 2007). As particles settle their modal radius grows and in the stratosphere, at heights of 30-40 km, it reaches 10-20 nanometers (Megner, 2007). According to calculations the largest part of meteoric aerosol is concentrated in this layer. The total number of meteoric particles in a column of the atmosphere with an area of 1 cm^2 can be estimated by means of integration of the curve in Fig. 1B in the height up to 90 km:

$$n_{tot} = \int_0^{90} n(z)dz = 1.0 \times 10^{11} \ cm^{-2} \quad (1)$$

Thus their total number in the all atmosphere of the Earth is:

$$N_{tot} = n \cdot S_{Earth} = 5.1 \times 10^{29} \quad (2)$$

In order to estimate the total mass of the meteoric substance present in the atmosphere, it is necessary to know the life time of a dust particle. The time of gravitational sedimentation of particles with a radius of a few nanometers from the height of 80 km to the tropospheric altitude is at least 150 years (Ogurtsov and Raspopov, 2011). In fact, the particles of meteoric smoke are unlikely to exist in the atmosphere such a long period – the circulating processes in mesosphere cause considerable vertical mixing of air masses. For example, Forkman et al., (2005) revealed that the vertical transfer of air may have speeds up to 250-450 m/day over the 65-90 km altitude. Balsley and Riddle,

(1984) found in the mesosphere a vertical wind with a velocity of 25 cm/s (21 km/day). Nevertheless, let us suppose that the life-time of the particle of the meteoric dust in atmosphere is ca 150 years. For the mass of the evaporated meteor substances, which are the source of the atmospheric aerosol of cosmic origin, the maximum estimate of 30 kt/yr (Lal and Jull, 2002) will be chosen. Then the total mass of extraterrestrial substance in the Earth's atmosphere will be M=30 kt/year ×150 years = 4.5 megatons (4.5×10^{12} g). This matter might influence terrestrial climate by different ways, first of all directly, by means of extinction (absorption and scattering) of the visible light. Cosmic dust produces in the atmosphere a layer of thickness:

$$\Delta l = \frac{4.5 \cdot 10^{12}\ g}{3\ g\ cm^{-3} \cdot 5.1 \cdot 10^{18}} = 2.9 \cdot 10^{-7}\ cm = 2.9\ nm. \tag{3}$$

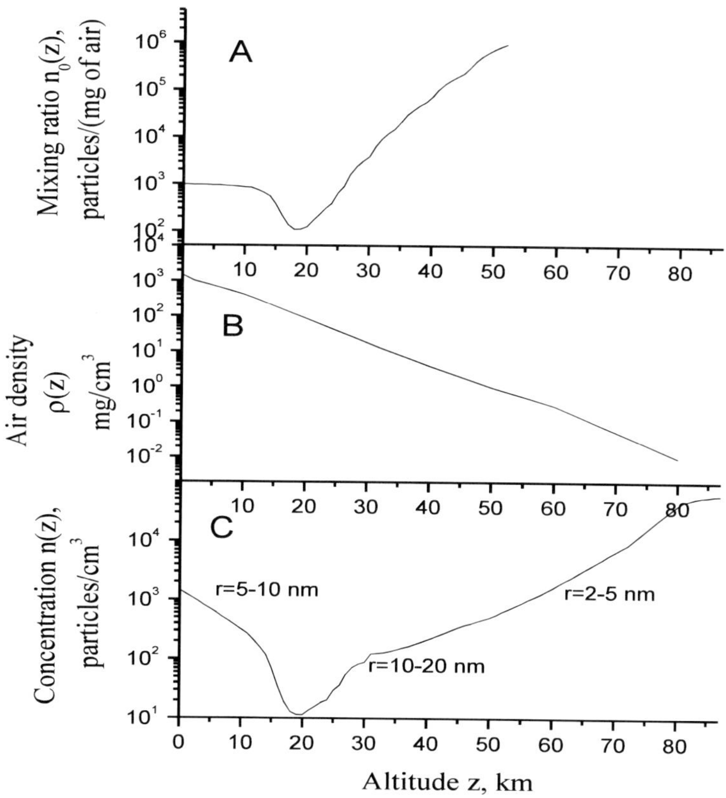

Figure 1. (a) Mixing ratio (number of meteoric aerosol particles per milligram of air), (b) air density, and (c) concentration of meteoric aerosol particles.

This value is about 200 times less than the mean wavelength of the visible light. Estimations of Muller and Mac Donald (2000) give even a lesser value of Δl ca 0.2 nm. Such a thin layer is not able to scatter sunlight efficiently and, therefore, the effect of atmospheric turbidity caused by extraterrestrial dust is negligible. Consequently a direct climatic influence of cosmic dust is insignificant.

However the extraterrestrial particles, which reach the stratospheric, and especially lower altitudes, might operate as CN and thus affect climate indirectly, via changes in aerosol concentration and cloudiness. Let us consider this possibility. The concentration of meteoric aerosol in the troposphere, calculated in work of Hunten rt al. (1980) is shown in Fig. 2 together with the data on the concentration of CN of all types taken from the works of Selezneva (1962), Ivlev (1969) and Matveev (1976).

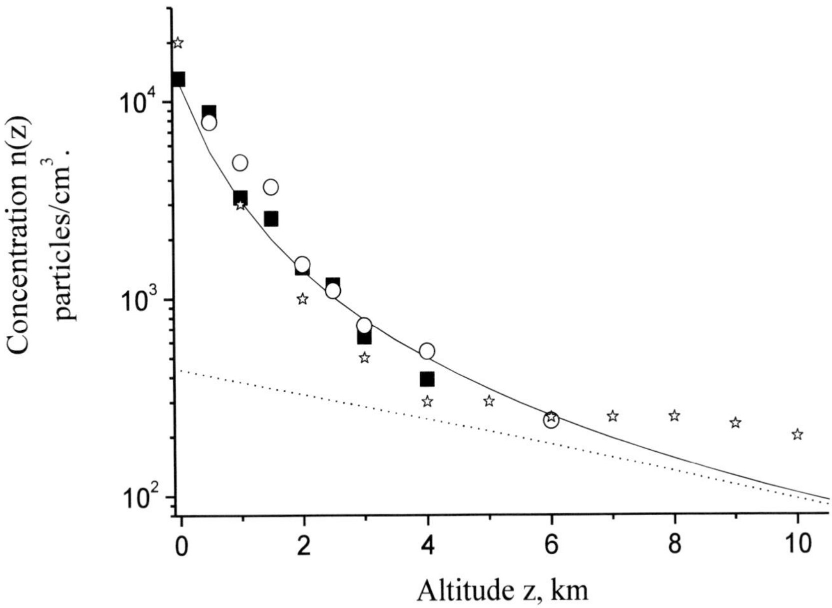

Figure 2. Concentration of different aerosols in the troposphere. The dashed line shows the concentration of cosmic particles (Hunten et al., 1980). Other lines show the concentration of condensation nuclei. The data of Ivlev (1969) is presented with asterisks, the data of Selezneva (1962) with open circles, the data of Matveev (1976) with squares, and approximation according to formula $n(z) = \dfrac{12500 \cdot 1.0}{(1.0 + z(km))^2}$ (Matveev, 1976) with a solid line.

Integration of the curves plotted in Fig. 2 shows that the total number of the CN in a column of troposphere with an area of 1 cm^2 can reach (1.2-1.6) $\times 10^9$, and the total number of particles of cosmic origin is 2.3×10^8. The effective radius of these particles is 5-10 nanometers, i.e. atleast in part they can serve as condensation nuclei. Therefore 15-20 % of the CN floating in the troposphere might be of cosmic origin. At the top of the troposphere cosmic particles may account for about 50 % of the CN. In the upper stratosphere (heights of 30-50 km), the aerosols of the meteoric nature most likely prevail (Megner, 2007).

Voigt et al., (2004) studied the polar stratospheric clouds and obtained evidence for an appreciable role of meteoric particles in their formation. The evidence of the influence of a meteoric dust on atmospheric precipitation has been shown in the work of Granitsky and Borisevich (2000). Thus, the hypothesis of Ermakov et al. (2006) that cosmic dust particles can affect the cloud cover seems to be fairly justified. Another possible mechanism of influence of extraterrestrial aerosol on climate may involve changes in planetary electric circuit. Since many particles of meteoric smoke are charged, they may influence the global electric current. The dependence of atmospheric conductivity on height, mainly deduced from the calculations, is shown in Fig. 3.

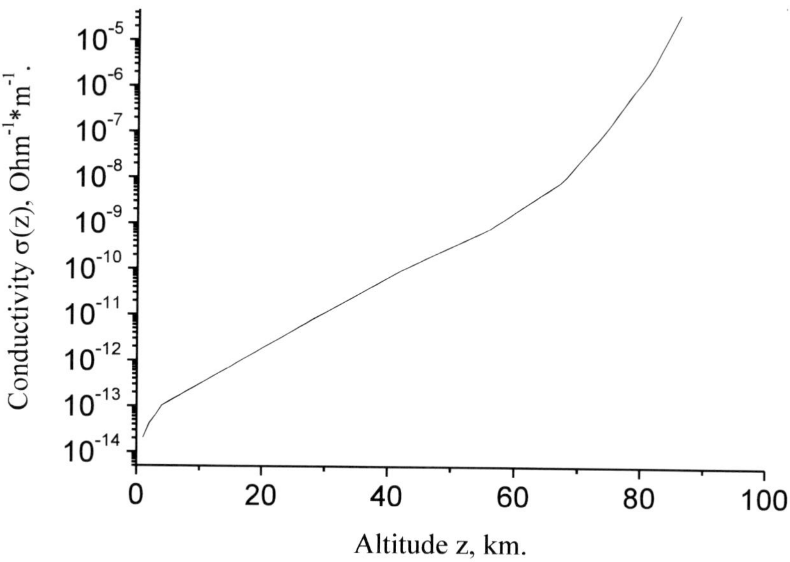

Figure 3. Altitude variations of the atmospheric conductivity σ (Reid, 1986; Rokityansky, 1981).

This conductivity is connected with atmospheric ions. Total resistance between the Earth's surface and the ionosphere can be estimated, by integration of the σ(z) curve over a height from 0 to 80 km. As a result we have $R_c=1.2\times10^{17}$ Om×m^2. The total resistance of the atmosphere is thus 230 Ohm. The mesospheric (heights of 50-80 km) resistance is about 0.04 Ohm. However, as reported by Zadorozhny and Tyutin (1998), Zadorozhny (2001), the considerable electric fields observed in mesosphere (up to 1 V/m) testify that the real effective conductivity in the middle atmosphere is 2-3 orders of magnitude lower than the ionic conductivity calculated theoretically and measured experimentally. The effect is probably connected with the influence of charged particles of meteoric origin settling in the mesosphere (Zadorozhny and Tyutin, 1998; Zadorozhny, 2001). Thus, the real resistance of the mesosphere might reach 4-40 Ohm i.e. more then 10 % of the total atmospheric resistance. Therefore the mesospheric resistance, depending on the cosmic dust flux, could influence the global current chain appreciably. This means that the meteoric substance entering the Earth's atmosphere can affect the atmospheric electric processes, which, in turn may have an effect on cloudiness (Tinsley et al., 2007).

Thus, the analysis performed above confirms that the flux of interplanetary cosmic dust in the Earth's atmosphere could influence global climatic processes indirectly. This highlights the value of further research on the cosmic dust-climate link. However, it was shown by Ogurtsov and Raspopov (2011) that the dust of interstellar origin has no impact (neither direct nor indirect) on the Earth's climate.

POSSIBLE INFLUENCE OF ASTEROID IMPACT ON ATMOSPHERIC PROCESSES

The collisions of giant (> 1 km) cosmic bodies with the Earth are expected to cause global catastrophes with prolonged consequences (Alvarez et al. (1980); Pope et al. (1994); Toon et al. (1997), Bailer-Jones (2009)). The impacts of mid-size bodies might also have appreciable effect on the Earth environment and climate, since they are more frequent.

The Hallstatt Cycle and Taurid Stream

The 2.0-2.5 kyr variation in radiocarbon concentration is reliably established (Houtermans, 1971; Damon and Sonett, 1991; Damon and Jiricowic, 1992; Vasiliev et al, 1999). Similar variation also is present in many climatic records (see Vasiliev et al (1999 and references therein). Damon and Jiricowic (1992) named this variation as Hallstatt cycle (HC), because of its apparent correlation with climate epochs documented by Schmidt and Gruhle (1988). The Hallstatt cycle is distinctly manifested in many paleoclimatic and archaeological data (Dansgaard et al., 1984; O'Brien et al., 1995), including the radiocarbon record obtained in framework of the INTCAL98 program (Stuiver et al., 1998). This time series, which in part cover the time interval prior to 11854 BP, is based on $\Delta^{14}C$ measurement in a number of mid-latitude Northern Hemisphere trees (Germany, Ireland, Washington, Oregon, California), and it is plotted in Fig. 4 after removing the long-term (T>5000 yr) trend.

Cosmogenic radiocarbon originates in the Earth's stratosphere and troposphere due to the effect of energetic galactic cosmic rays (GCR). Since in the inner heliosphere the low energy ($<10^3$ GeV) GCR are modulated by solar activity, the changes of the concentration of ^{14}C in atmosphere reflects the corresponding solar variations and thus radiocarbon could serve as solar proxy. The Hallstatt cycle is evidently expressed in the smoothed $\Delta^{14}C$ data and particularly in the 1800 year wavelet component (see Figure 4b). The origin of this 2.0-2.5 kyr periodicity still is unknown. Hood and Jirikowic, (1990); Vasiliev et al, (1999) presumed that this variation is of solar origin. According to their assumption the HC influences the atmospheric radiocarbon by two ways: (a) directly, via solar modulation of the intensity of GCR; (b) indirectly, via solar-driven changes of global climate, which, in turn, cause the corresponding variations of parameters of the carbon exchange system. In order to test this hypothesis we compared the radiocarbon record with the cosmogenic ^{10}Be time series. The radioberyllium ^{10}Be also is a proxy of solar activity, but it is less dependent on the global-scale climatic processes than ^{14}C. Its concentration has been measured in central Greenland ice core over the time intervals 3300-8000 BP and 9400-40000 BP in the framework of project GISP2 (Alley et al., 1995; Finkel and Hishiizumi, 1997). Since GISP2 and INTCAL98 data sets overlap partly we can compare the Hallstatt periodicities in both series. This comparison is illustrated in Fig. 5.

The dates of intersection of the Earth's orbit by core of Taurid meteoroid stream are marked with arrows.

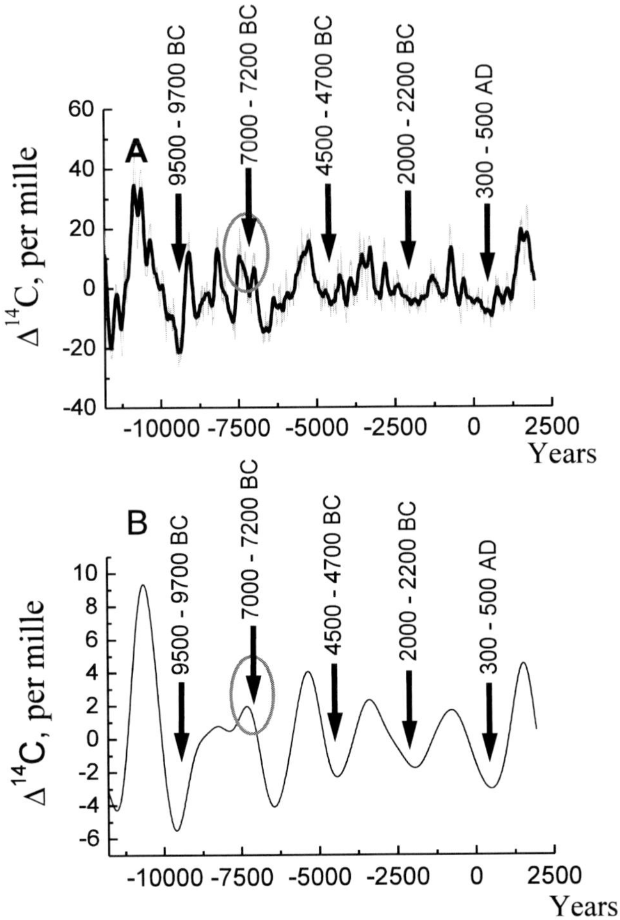

Figure 4. A – residual part of the atmospheric ^{14}C profile (thin grey line - raw data, thick black line – data smoothed by 25 points), B – 1800 year wavelet component (MHAT basis) of the residual part.

It is evident from Fig. 5 that despite the similarity in long-term (the periods more than 600 years) variations in ^{10}Be and $\Delta^{14}C$ records, they also have appreciable differences: (a) concentration of ^{10}Be has no fall ca. ~9600 BC unlike $\Delta^{14}C$, (b) – the HC in beryllium through 1300-6000 BC is distorted by more short-term fluctuations. They provide spikes in the concentration of ^{10}Be ca 2150 and 4500 BC. These peaks are absent in $\Delta^{14}C$. These divergences allow for the assumption that some other factors contribute to the ~2400-year periodicity of radiocarbon besides the Sun's activity.

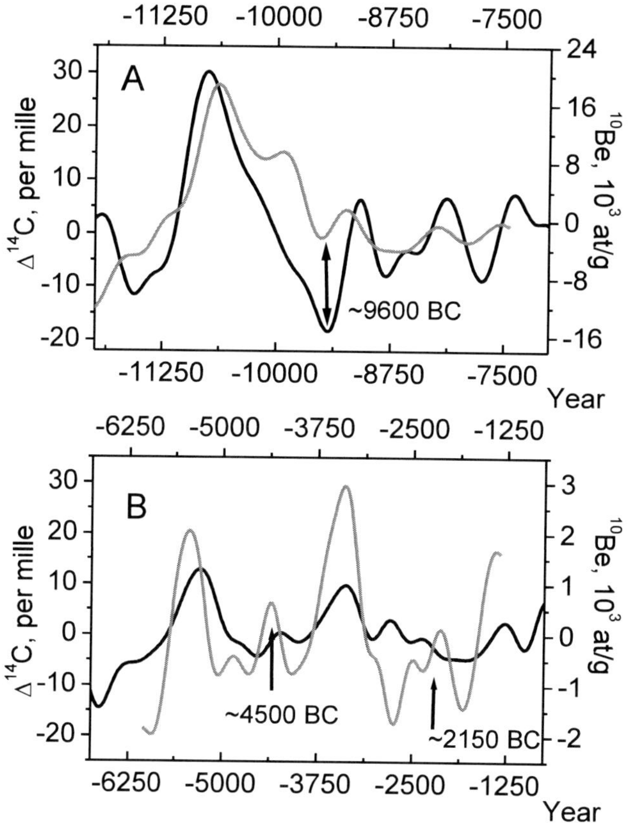

Figure 5. Beryllium and radiocarbon records filtered by Fourier-filter (time variations with periods more than 600 year subtracted). Black line – concentration of ^{14}C in tree rings (Stuiver et al., 1998), gray line – concentration of ^{10}Be in Greenland ice (Finkel and Nishiizumi, 1997).

Thereupon it is pertinent to note a hypothesis about the nodal precession of the core of the Taurid cometary-meteoroid complex, which has been stated in works of Asher and Clube (1993, 1997) and Clube (1998). Taurids is the broad stream of cometary debris, which the Earth passes through each year during June-July and September-December. This meteoroid complex was likely formed during the last 10-20 thousands of years as a result of the defragmentation of a giant comet (Steel, Asher, Clube, 1991; Steel and Asher, 1996). At the heart of this stream lies a narrower core whose orbital nodes are assumed to intersect the terrestrial orbit every 2.25-2.5 Kyr during a timespan of about two centuries (Asher and Clube, 1993; Clube, 1998). Obviously, the bombardment of the Earth by cometary debris is much more intensive during these intersection periods. The periods of nodal precession of the Taurid Complex and of Hallstatt climatic cycle are very similar. Moreover, extremes of 2.5 kyr climatic variation coincide well with the epochs of nodal intersections. The last two pairs of such events (about 1500 and 4000 years ago) were determined reliably by Asher and Clube (1993) and Clube (1998). In Figure 4 we continued intersection dates backwards in time to 10 kyr. The good agreement between these dates and minima of long-term changes of radiocarbon is apparent. That is why Clube (1998) assumed that HC is a cycle of global coolings of the Earth ("impact winter") as a result from the input of tremendous volume of dust and soot into the atmosphere due to asteroid downfalls. However, in order to cause a global effect, the energy of impact must be more than 2×10^5 Mt TNT (Medvedev et al., 1996). Such energy release can arise from the incidence of a stony asteroid with a diameter of 1.5 km and velocity 20 km/s. The probability of random collision between Earth and such asteroid is very small – less than one per 10^5 years (Medvedev *et al.*, 1996). Naturally this probability should increase sharply when the Earth intersects the core of Taurid complex. However, it is difficult to answer whether or not this probability can rise up to 40-50-fold. Fortunately pertinent paleoclimatic data allows a look at the past of the Hallstatt climatic cycle in a novel fashion.

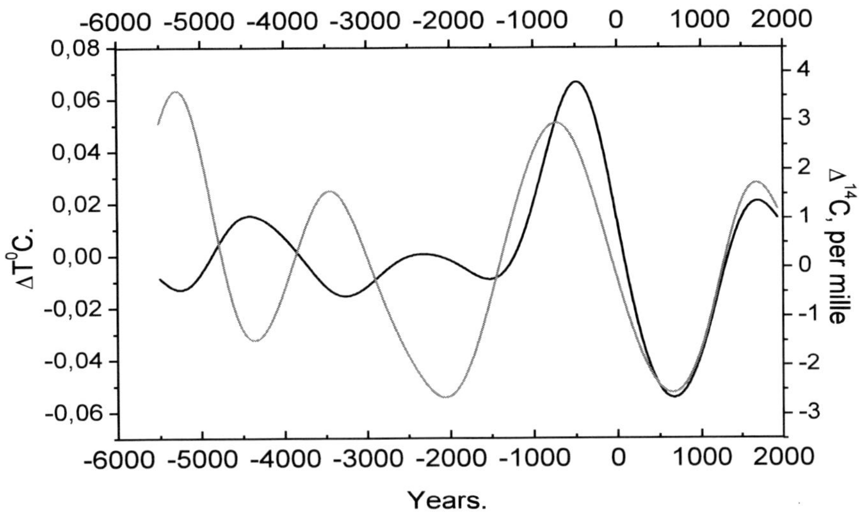

Figure 6. Wavelet filtered (1580-2700 scale band, MHAT basis) series of atmospheric radiocarbon concentration (grey line) and northern Fennoscandian July temperature (black line).

Approximately 2.1 kyr variation was revealed in a long annually resolved reconstruction of July temperature in northern Fennoscandia (~67^0N) made by Lindholm and Eronen (2001). Radiocarbon and northern Fennoscandian (NF) temperature records, wavelet filtered in 1580-2700 yr scale band, are plotted in Figure 6.

Figure 6 shows that the phase relationship between two series is not constant – they correlate positively after ca. 1500 BC and negatively before that. It means that after 1500 BC the maxima of 2.4 kyr variation of atmospheric ^{14}C occur together with the maxima of corresponding cycle of NF regional temperature. Little Ice Age (AD ~1450-1800), often considered as a period of global cooling and clearly marked in radiocarbon record by substantial increase of $\Delta^{14}C$, in NF was a period of higher summer temperature. Not only the NF proxy does lack a manifestation of the Little Ice Age. Networks of annually-resolved time series from trees, sediments, ice cores, corals, and historical documents are emerging, which support the idea that globally synchronous cold periods longer than a decade or two did not occur within the last 500 years (see Overpeck (1995) and references therein). Thus, it is possible that the significant regional and spatial variability of climate took place in the past millennium (Overpeck, 1995). Taking that into account one can assume that HC is not a cycle of global terrestrial cooling and warming, but has spatial and probably seasonal variation, too. If it is so, the HC is likely not an oscillation of the total energy contained in the lower troposphere but a variation in distribution of energy between different regions of our planet. Such variations in regional distribution of heat may arise as a result of changes in atmospheric circulation. Atmospheric dynamics is very sensitive to changes in a chemical composition of the atmosphere, particularly to variations in ozone and nitrogen oxides concentration (Pudovkin and Raspopov, 1992; Pudovkin, 1996; Shindell, 1999; Haigh, 1996). Hence, even a strong forcing as 10^5 MT impact of asteroid might not be absolutely necessary to change the pattern of the global atmospheric circulation. Let us estimate the consequences of the Earth's collision with a medium size asteroid. Firstly it should be noted that the asteroid is likely to fall into the ocean, because water covers more than 70% of terrestrial surface. We will consider an asteroid with a diameter of 250 meters for further evaluation. Meteoroid of such a size can impact the Earth once per ca 6 kyr if collisions are randomly distributed. However, as we have stated above the Earth's orbit is occasionally intersected by the core of Taurid Stream, which increases the probability of impact strongly. Thus, it is not too unrealistic to imagine collision with a 250 meter asteroid every time when the Earth's orbit is within the core of Taurid Complex (once per 2.5 kyr).

Possible Climatic Consequences of Impact of a Stony Asteroid with a Diameter of 250 M

The energy of impact with the stony asteroid (ρ = 3.0 g×sm^{-3}) of a medium size (R = 125 m) moving with velocity of v=25 km×s^{-1}, can be calculated by formula:

$$E_{imp} = \frac{4}{3}\pi R^3 \cdot \frac{\rho v^2}{2} \cong 1800 Mm \quad THT \tag{4}$$

Such energy release can disturb oceanic circulation, increase mixing between deep and surface water layers and enrich the surface layer by deep water. Because CO_2 in deep ocean is depleted by ^{14}C, this process can lead to a corresponding decrease in atmospheric radiocarbon concentration. However, the volume of increased deep-surface mixture will be small (1 Gt of energy is able to heat by 1^0C only 10^3 cubic kilometres of water) and, hence, a global effect will likely be weak. Climatic consequences can be more significant. It has been shown that falling of an asteroid of medium (200-300 m) size in the ocean would cause large-scale perturbation of upper atmospheric chemistry, including destruction of an appreciable part of the ozone layer (Klumov, 1999; Birks et al., 2005). If an asteroid of 500 m – 2 km size falls in equatorial part of the Earth, ozone depletion could be of a global scale (Pierazzo et al., 2011). A simple estimation can illustrate the effect of a meteoroid fall on the ozone layer. If impactor (ρ = 3.0 g×sm^{-3}, R = 125 m, v=25 km×s^{-1}) strikes the Earth's ocean, a cavity in diameter about 5 km and depth about 3 km (Crawford and Mader, 1998) is formed. The volume of this cavity will reach approximately 50 км^3. Considerable part of the water, occupying the given cavity, will get to the atmosphere in the form of splashes and steam. According to the work of Toon et al. (1994), the mass of water which reach the atmosphere in the form of steam, is about 11 masses of the projectile, i.e. 2.1×10^3 Mt of water. The atmosphere will get about 4×10^3 Mt of water in the form of splashes (Toon et al., 1994). It is easy to assess the height in the atmosphere which this water volume can arrive at:

$$h = \frac{E_{ok}}{m_{ok} g}. \qquad (5)$$

Where m_{ok} – the mass of the water ejected into the atmosphere, E_{ok} - the energy transferred to this water bulk, h - average height of lifting of the water volume in the atmosphere. Thus, if only 3 % of the total kinetic energy of impact will be converted to E_{ok}, the average height of the water bulk loft reach 40 km. It is obvious that a large part of the ejected sea water can reach the stratospheric and probably mesospheric altitude. The splashed liquid water will be removed quickly (probably within several days) and, therefore, will unlikely influence the atmosphere chemistry. Water in the form of steam (2.1×10^3 Mt) can live in the stratosphere for several years (Toon et al., 1994) and, hence, is chemically active. Thermodynamical estimations show that the vaporized water contains ca 1% of NaCl and HCl and ca 0.01% of Cl (Klumov, 1999). That means that the total mass of chlorine, instantly delivered to the middle atmosphere by the impact of a meteoroid, will be about 1.5% of the water mass, i.e. 30 Mt. This value corresponds to 5×10^{35} atoms of Cl. In comparison all modern industry produce about 1 Mt chlorine per year as input into thestratosphere (Prather *et al*, 1990). Hence, the collision of an asteroid with the ocean will inject into the stratosphere an amount of reactive chlorine equal to that produced by all modern industry during 30 years. The role of chlorine in ozone loss is well known. It breaks O_3 molecules by means of reactions:

$$Cl + O_3 \rightarrow ClO + O_2 \qquad (6)$$

$$ClO + O \rightarrow Cl + O_2 .$$

This way one atom of chlorine can destroy 10^5 molecules of ozone. Taking into account that the total number of O_3 molecules in the atmosphere is about 4×10^{37}, the possible damage to the ozone layer is evident. Another effective destroyer of ozone is nitrogen oxide NO. It is formed in the equilibrium reaction $N_2+O_2\leftrightarrow 2NO$ proceeding under high (more than $1500\text{-}2000^0$ C) temperatures in the fireball forming in the atmosphere by falling meteoroids. According to Pittock et al. (1986) 1 Mt of allocated energy forms 10^{32} molecules of NO, i.e. explosion with energy more than 10^3 Mt will produce more than 10^{35} NO molecules. It is very likely that the falling of asteroid will cause also some other changes in the stratospheric chemistry which are difficult to predict. Nevertheless, only Cl and NO, created due to the impact, might change the chemical composition of the stratosphere significantly. The estimation of Klumov (1999) shows that the impact of mid-size asteroid can create an ozone hole with a size of a few thousands of kilometers. The model calculations of Birks et al. (2007) show that at altitudes more than 40-45 km, globally averaged ozone depletion can exceed 1.5% after a month following the impact. The time interval, which is necessary to stratospheric ozone to recover, could reach a few months (Klumov, 1999). That means that if the asteroid of 250 m diameter strikes the Earth's ocean it can appreciably change the energy balance of the stratosphere, which in turn, would transform the atmospheric circulation pattern considerably. Indeed, a few percent variation of the concentration of stratospheric ozone during the 11-year solar cycle is enough to change the atmospheric circulation appreciably (Haigh, 1994, 1996; Shindell et al., 1999; Tourpali et al., 2003; Rozanov et al., 2004). The sharp and instant change of energy balance of the atmosphere might serve as a trigger, which switch up climatic processes with a much longer time scale.

Of course, it is impossible to explain a 2400-year climatic cycle only by periodic fallings of asteroids belonging to the Taurid complex. However, the chain of such events coinciding with extremes of the HC (300-500 AD, 2000-2200 BC, 4500-4700 BC, 9500-9700 BC, see Fig. 4) can resound with the periodicity, strengthen it and serve as a rhythm-driver. Thereupon, it is worth to draw attention to the time interval 7000-7200 BC. This is the only epoch of intersection of the Earth's orbit by the Taurid stream core, which coincides with the maximum (not with the minimum) of the HC in $\Delta^{14}C$. Fig. 4B shows that in spite of the long-term increase of the radiocarbon concentration, linked to the HC maximum, just during 7000-7200 BC the curve of $\Delta^{14}C$ has a distinct drop.

The analyses performed in the work support the idea that the downfall of a stony asteroid with a density of 3.0 g×cm^{-3}, velocity 25 km×s^{-1} and 250 m in diameter, which occur every 2.5 kyr, might contribute to the quasi 2.4 kyr earthly climatic cycle. Obviously, this hypothesis, based on the rather arbitrary assumptions, needs further verification and additional testing:

1) to evaluate more reliably and accurately the probability of collision between the Earth and the asteroids of Taurid complex, particularly during the intersection periods.
2) to estimate more reliably and accurately the probable effect of asteroid impact on the Earth's atmosphere, including its possible duration.

Building and updating long and reliable climatic proxies, which would allow tracing the spatial and seasonal distribution of the Hallstatt variation, are also desirable future goals.

Conclusion

The amount of evidence of the possible influence of meteorically derived substances on climate processes is increasing. That means that the flux of the extraterrestrial substances (both dust particles and celestial bodies of an appreciable size), may potentially prove be one of the climate-forcing agents. The details of these chains of reasoning merit further investigation, viz. linking climatic processes and cosmic matter entering the atmosphere as well as increasing our knowledge of the temporal and spatial distribution of the extraterrestrial accretion.

References

Alley, R.B., Finkel, R.C., Nishiizumi, K. et al. (1995). Changes in continental and sea-salt atmospheric loadings in central Greenland during the most recent deglaciation: Model-based estimates. *J. Glaciol.* Vol. 41, pp. 503-514

Alvarez, L. W., Alvarez, W., Asaro, F. and Michel, H. V. (1980). Extraterrestrial cause for the cretaceous-tertiary extinction. *Science*. Vol. 208, pp. 1095-1108

Asher, D.J. and Clube, S.V.M. (1993). An extraterrestrial influence during the current glacial-interglacial. *Quat. Journ. Roy. Astr. Soc.* Vol. 34, pp. 481-511

Asher, D.J. and Clube, S.V.M. (1997). Towards a dynamical history of "proto-Encke". *Celest. Mech. and Dynam. Astron*. Vol. 69, pp. 149-170.

Bailer-Jones, C. (2009). The evidence for and against astronomical impacts on climate change and mass extinctions: A review. *Int. J. Astrobiology*. Vol. 8, pp. 213-219

Balsley, B. B. and Riddle, A. C. (1984). Monthly mean values of the mesospheric wind field over Poker Flat, Alaska . *J. Atmos. Sci.* Vol. 41, pp. 2368-2375

Birks, J.W., Crutzen P.J. and Roble G. R. (2007). Frequent ozone depletion resulting from impacts of asteroids and comets. In Bobrowsky P., Rickman H. (Eds.), *Comet/Asteroid Impacts and Human Society* (pp. 225-254). Springer Verlag

Ceplecha, Z. (1996). Luminous efficiency based on photographic observations of the Lost-City fireball and implications for the influx of interplanetary bodies onto Earth. *Astron. Astrophys.* Vol. 311, pp. 329-332

Clube, S.V.M. (1998). The problem of historical catastrophism. In B.J. Peiser, T. Palmer and M.E. Bailey (Eds.), *Natural catastrophes during Bronze Age civilizations: archaecological, geological, astronomical and cultural perspectives* (p. 242-251). Oxford: Publ. Archaeopress

Crawford, D.A. and Mader, C.L. (1998). Modelling asteroid impact and tsunami. *Sci. Tsunami Hazards*. Vol. 10, pp. 21-31

Damon, P.E. and Sonett, C.P. (1992). Solar and terrestrial components of the atmospheric ^{14}C variation spectrum. In C.P. Sonett, M.S. Giampapa, M.S. Mathews (Eds), *The Sun in time* (pp. 360-368). Tuscon: Univ. of Arizona press

Damon P.E. and Jirikowic, J.L. (1992). Solar forcing of global climatic change? In R.E. Taylor, A. Long, R.S. Kra Eds), *Radiocarbon after four decades: an interdisciplinary perspective* (pp. 117-129). New York: Springer-Werlag

Dansgaard, W., S. J. Johnsen, H. B. Clausen, D. Dahl-Jensen, N. Gundestrup, C. U. Hammer, and H. Oeschger. (1984). North Atlantic climate oscillations revealed by deep Greenland ice cores. In J.E. Hanson and T. Takahashi (Eds.), *Climate Processes and Climate Sensitivity, Geophysical Monograph Series* (Vol. 29, pp. 288-298). Washington: American Geophysical Union

Dohnanyi, J.S. (1972). Interplanetary objects in review: statistics of their masses and dynamics. *Icarus*. Vol. 17, pp. 1-48

Ermakov V.I., Okhlopkov V.P. and Stozhkov Yu.I (2009). Influence of cosmic rays and cosmic dust on the atmosphere and Earth's climate. Bull. *Russian Academy Sciences: Physics*. Vol. 73, pp. 416–418

Esser, B.K. and Turekian, K.K. (1988). Accretion rate of extraterrestrial particles determined from osmium isotope systematics of Pacific pelagic clay and manganese nodules. *Geochim. Cosmochim. Acta*. Vol. 52, pp. 1383-1388.

Finkel, R.C. and Nishiizumi, K. (1997). Beryllium 10 concentrations in the Greenland Ice Sheet Project 2 ice core from 3-40 ka. *J. Geophys. Res.* V. 102, pp. 26699-26706

Forkman, P., Eriksson, P., Murtagh, D. and Espy, P. (2005). Observing the vertical branch of the mesospheric circulation at latitude 60°N using ground-based measurements of CO and H_2O. *J. Geophys. Res.* Vol. 110, D05107, doi:10.1029/2004JD004916

Granitsky L. V. and Borisevich, A. N. (2000). Research of influence of the meteoric stream on the weather condition : preliminary consideration. In *Proceedings of SPIE, the International Society for Optical Engineering* (pp. 563-570). V. 4341

Haigh, J.D. (1996). The role of stratospheric ozone in modulating the solar radiative forcing of climate. *Nature*. Vol. 370, pp. 544-546

Hood, L.L. and Jirikowic, J.L. (1990). Recurring variations of probable solar origin in the atmospheric $\Delta^{14}C$ time record. *Geophys. Res. Lett.*Vol. 17, pp. 85-92

Houtermans, J.C. (1971). *Geophysical interpretation of Bristlecone pine radiocarbon measurements using a method of Fourier analysis of unequally spaced data.* Ph.D. Thesis. University of Bern.

Hunten, D.M., Turco, R.P. and Toon, O.B. (1980). Smoke and dust particles of meteoric origin in the mesosphere and stratosphere. *J. Atmosph. Sci.* Vol. 37, pp. 1342–1357

Ivlev, L.S. (1969). Aerosol model of the atmosphere. In *Problems of atmospheric physics* (pp. 125-160), Leningrad: LGU press (in Russian)

Kane, T. J. and Gardner, C. S. (1993). Lidar Observations of the meteoric deposition of mesospheric metals. *Science*. Vol. 259, pp. 1297-1300

Kasatkina, E.A., Shumilov, O.I., Lukina N.V. et al. (2007). Stardust components in tree rings. *Dendrochronologia*. Vol. 24, pp. 1311-135

Klumov, B. A. (1999). Destruction of the ozone layer as a result of a meteoroid falling into the ocean. *J. Exp. Theor. Phys. Lett.* Vol. 70, pp. 363-370

Kyte, F. T. and Wasson, J. T. (1986). Accretion rate of extraterrestrial matter: Iridium deposited 33 to 67 million years ago. *Science*. Vol. 232, pp. 1225-1229

Lal, D. and Jull, A.J.T. (2002). Atmospheric cosmic dust fluxes in the size range 10^{-4} to 10 centimeters. *Astrophys. J.* Vol. 576, pp. 1090-1097

Lindholm, M and Eronen, M. (2000). A reconstruction of mid-summer temperatures from ring-widths of Scots pine since AD 50 in northern Fennoscandia. *Geografiska Annaler*. Vol. 82, pp. 527-535

Love, S.G. and Brownlee, D.E. (1993). A direct measurement of the terrestrial mass accretion rate of cosmic dust. *Science*. V. 262, pp. 550-553

Maurette, M., Jehanno, C., Robin, E. and Hammer, C. U. (1987). Characteristics and mass distribution of extraterrestrial dust from the Greenland ice cap. *Nature.* Vol. 328, pp. 699–702

Matveev, L.T. (1976). Course of the general meteorology. *Leningrad: Hydrometeoizdat* (in Russian)

Medvedev Yu.D., Sveshnikov M.L., Sokolsky A.G. et al. (1996). *Asteroid-comet hazard.* St. Petersburg: ITA RAS (in Russian)

Megner, L. (2007). Funneling of meteoric material into the polar winter vortex. In G.D. O'Dowd, P.E. Wagner (Eds), *Nucleation and atmospheric aerosols, 17th Int. Conf.* (pp. 860-864), Galway, Ireland

Muller, R.A. and Mac Donald, G.J. (2000). Ice ages and astronomical cause. *Data, spectral analysis and mechanisms.* London: Springer-Praxis

Ogurtsov, M.G. and Raspopov, O.M. (2011). Possible impact of interplanetary and interstellar dust fluxes on the Earth 's climate. *Geomagn. Aeronomy.* Vol.51, pp.275 –283

Overpeck, J.T, (1995). Paleoclimatology and climate system dynamics. *Revi. Geophys.* Vol. 33, pp. 863-871

Pierazzo, E., Garcia, R.R., Kinnison, D.E. et al. (2010). Ozone perturbation from medium-size asteroid impacts in the ocean. *Earth Planet. Sci. Lett.* Vol. 299, pp. 263-273

Pittock A.B., Ackerman T.P., Crutzen P.J. et al. (1986). *Environmental consequences of nuclear war.* Chichester-New York-Brisbane-Toronto-Singapore: J. Wiley and Sons.

Pope, K.O., Baines, K.H., Ocampo, A.C. and Ivanov, B.A. (1994). Impact winter and the Cretaceous/Tertiary extinctions: Results of a Chicxulub asteroid impact model. *Earth Planet. Sci. Lett.* Vol. 128 pp. 719-725

Prather, M.J., Garcia, M.M., Douglass, A.R. et al. (1990). The space Shuttle's impact on the stratosphere. *J. Geophys* Res. Vol. 95, pp. 18583-18590

Pudovkin, M.I. and Raspopov, O.M. (1992). Mechanism of the influence of solar activity on the state of lower atmosphere and meteoparameters. *Geomagn. Aeron.* Vol. 32, pp. 1-22,

Pudovkin M.I. (1996). Effect of solar activity on the lower atmosphere and weather. *Soros Education J.* Vol. 10, pp. 106-114 (in Russian)

Rapp, M., Strelnikova, I. and Gumbel, J. (2007). Meteoric smoke particles: Evidence from rocket and radar techniques. *Adv. Space Res.* Vol. 40, pp. 809–817

Rasmussen, K.L., Clausen, H.B. and Kallemeyn, G.W. (1995). No iridium anomaly after the 1908 Tunguska impact: evidence from a Greenland ice core. *Meteoritics.* Vol. 30, P. 634–638

Reid, G.C. (1986). Electrical structure of the middle atmosphere. In *The Earth's electrical environment*. National Academy Press, Washington D.C.

Rokityansky, I.I. (1981). *Inductive sounding of the Earth*. Kiev: Naukova Dumka (in Russian)

Rosinski, J. and Snow, R.H. (1961). Secondary particulate matter from meteor vapors. *J. Meteorology*. Vol.18, pp. 736–745

Rozanov, E.V., Schlesinger, M.E., Egorova, T.A. et al. (2004). Atmospheric response to the observed increase of solar UV radiation from solar minimum to solar maximum simulated by the University of Illinois at Urbana-Champaign climate-chemistry model. *J. Geophys. Res.* Vol. 109, D01110, doi:10.1029/2003JD003796

Selezneva, E.S. (1962). Peculiarities of the vertical distribution of condensation nuclei at different atmospheric stratifications. *Proc. GGO*. Vol. 134 (in Russian)

Schmidt, B. and Gruhle, W. (1988). Radiokohlenstoffgehalt und dendrochronologie. *Naturwissenshaftliche Rundschau*. Vol. 5, pp. 177-182

Shindell, D., Rind, D., Balachandran, N. et al. (1999). Solar cycle variability, ozone and climate. *Science*. Vol. 284, pp. 305-308.

Steel ,D.I., Asher, D.J. and Clube, S.W.M. (1991). The structure and evolution of the Taurid complex. Month. *Not of Roy. Astr. Soc.* Vol. 251, pp. 806-822

Steel, D.I. and Asher, D.J. (1996). The orbital dispersion of the macroscopic Taurid objects. Month. *Not of Roy. Astr. Soc.* Vol. 280, pp. 806-822

Stuiver, M., Reimer, P. J., Bard, E. et al. (1998). INTCAL98 radiocarbon age calibration, 24,000-0 cal BP. *Radiocarbon*. Vol. 40, pp. 1041-1083

Taylor S., Lever J. H. and Harvey R. (1998). Accretion rate of cosmic spherules measured at the South Pole. *Nature*. Vol. 392, pp. 899–903

Tinsley, B. A., Burn, G. B. and Zhou, L. (2007). The role of the global electric circuit in solar and internal forcing of clouds and climate. *Adv. Space Res.* Vol. 40, pp. 1126–1139

Tirsky, G.A., (2000). Interaction between cosmic bodies and the Earth's and planet's atmospheres, *Soros Education J.* Vol. 6, pp. 76–82 (in Russian)

Toon, O.B., Zahle, K., Turco, R.P. and Cowey, C. (1994). Environmental perturbations caused by impacts. In Gehers T. (Ed.), *Hazards due to comets and asteroids* (pp. 791-827). Tuscon and London: The Univ. of Arizona Press

Toon, O., Morrision, D., Turco, R.P. and Covey, C. (1997). Environmental perturbations caused by the impacts of asteroids and comets. *Rev. Geophys.* Vol. 35, pp. 41-78

Tourpali, K., Schuurmarns, C.J.E., van Donald, R. et al. (2004). Stratospheric and tropospheric response to enhanced solar UV radiation: a model study. *Geophys. Res. Lett.* Vol. 30, DOI:10.1029/2002GL016650.

Vasiliev, S. S., Dergachev, V. A. and Raspopov, O. M. (1999). Sources of long-term variations in the radiocarbon concentration in the Earth's atmosphere. *Geomagn. Aeronomy.* Vol. 39, pp. 749-757

Voigt, C., Schlager, H., Luo, B.P. et al. (2004). Nitric acid trihidrate (NAT) formation at low NAT supersaturation in polar stratospheric clouds (PSCs). *Atmos. Chem. Phys.* Vol. 5, pp. 1371–1380

Zadorozhny, A.M. and Tyutin, A.A. (1998). Effects of geomagnetic activity on the mesospheric electric fields. *Annales Geophysicae.* Vol. 16, pp. 1544-1551

Zadorozhny, A.M. (2001). Effects of charged dust on mesospheric electrical structure. *Advances Space Res.* Vol. 28, pp. 1059-1064

In: Meteorites and Asteroids
Editors: A. Dementieva et al.
ISBN: 978-1-61942-115-8

Chapter 3

THE DIVERSION AND EXPLOITATION OF ICE-RICH NEOS USING THE SOLAR COLLECTOR

Gregory L. Matloff[*]
Physics Dept., New York City College of Technology, CUNY, Brooklyn, NY, US
Hayden Associate, American Museum of Natural History, New York, NY, US

ABSTRACT

A significant fraction of the Near-Earth Object (NEO) population may be objects that originated as comets. Kinetic or nuclear defection schemes may result in the fragmentation of an extinct comet, rather than its deflection. The Solar Collector (SC) has been proposed as a method of deflecting water-ice-rich NEOs on Earth-threatening trajectories. The SC, a two-sail solar sail, functions by concentrating sunlight against the NEO regolith to raise an energized jet that would alter the NEO's solar orbit as a consequence of Newton's Third Law. Recent work reported here expands upon a NASA-supported 2007 study. A dimensionally correct model of SC function is presented with a discussion of energy-loss mechanisms. Measurements of the electromagnetic penetration depth in an extraterrestrial sample are incorporated. The SC-NEO separation is

[*] E-mail: (GMatloff@citytech.cuny.edu).

maintained at about 1-km to reduce predicted SC-optics degradation from the energized jet. It is shown that a ~140-m radius SC can divert an ~200-m radius Earth-threatening NEO with a rotational period of about 5 $X10^4$ seconds if the SC remains on station for a period of years. Much smaller ice-rich NEOs in appropriate solar orbits could be diverted into high-Earth-orbits for resource utilization using an SC that remains on station for a sufficiently long time interval.

INTRODUCTION: THE NEO THREAT, THE NEO OPPORTUNITY

There are many natural and human-caused threats to our emerging global civilization—climate change, thermonuclear war, super volcanoes, Earth-quakes, tsunamis, nearby supernova, etc. Many of them can be neither predicted nor controlled. But at least one of these can potentially be mediated by our 21st deep-space technology—the threat from impacts by Near Earth Objects (NEOs). Originating from deep space (the Kuiper Belt at 30-50 Astronomical Units or AU from the Sun and the Main Asteroid Belt at 2-5 AU from the Sun), there may be hundreds of thousands of NEOs in the size range 20 m-10 km. As of May 2008, the NASA Jet Propulsion Laboratory NEO website tabulates orbital and known physical characteristics for about 8,000 NEOs [1]. About 900 of these are less than 30 m in (equivalent spherical) diameter, 1,700 range 30-100 m, 1,900 are between 100 and 300 m in size, 2,500 are 300 m-1 km in size and roughly 700 exceed 1 km.

What little we know about the physical characteristic of these objects has been learned from meteorite impacts, remote radar ranging and a few visits by space probes. But we do know that impacts by large celestial objects (such as the ~10-km one that contributed to the dinosaurs' demise about 65 million years ago) have contributed to mass extinctions of terrestrial life.

Large (>1km) objects seem to impact Earth at intervals of tens of millions of years. None of the larger NEOs in the JPL tabulation pose an immediate threat to the Earth. But smaller objects impact the Earth at much more frequent intervals. In 1908, an object in the 30-50 m range vaporized in the atmosphere above Tunguska, Siberia. The explosive yield of this event has been calculated as equivalent to that of a multi-megaton thermonuclear weapon. If the Tunguska object had "targeted" a city, millions may have been killed. Such "city-killing" NEOs may impact our planet at intervals of a few centuries [2].

In 2007, the NASA Marshall Space Flight Center in Huntsville, AL, conducted a study to investigate methods of diverting a medium-sized NEO on an Earth-impact trajectory. The timing of this study was influenced by the interest of NASA and other space agencies in developing a deep-space capability and the close encounters of NEO Apophis 99942 predicted for 2029 and 2036. This author was privileged to participate in this effort in the analysis of the Solar Collector (SC), a method of diverting an Earth-threatening NEO based upon the solar sail.

The premise of the 2007 NASA Marshall NEO study is that by about 2020, a heavy-lift rocket would be available and capable of launching ~10^5 kg on an Earth-escape trajectory [3]. Assuming such a capability, could a reliable NEO-intercept plan be implemented if predicted NEO impact date is limited to a decade or so in the future?

Unfortunately, none of the existing or projected NEO diversion strategies is perfect and could be applied to all types of NEOs. The favorite of Hollywood special- effects experts, the nuclear option has several drawbacks. A primary issue is the socio-political one. A ~10^5 kg payload of thermonuclear "devices" might translate as the equivalent of ~100 one-megaton-yield bombs in one rocket. Would various world powers trust a competing national space program preparing to launch such a device towards a distant NEO? Although nuclear explosives would likely divert a metallic NEO, a loosely held extinct comet nucleus or rubble pile might likely be fragmented or "calved" by the blast. Earth would then face impact by several NEO fragments instead of one larger NEO—and the fragments would be highly radioactive.

A second option is kinetic impact. In July 2005, the 400 kg NASA Deep Impact probe slammed into the nucleus of Comet Tempel 1 at a relative velocity of about 10 km/s. A crater with dimensions of several hundred meters was created and a visible fireball was raised by the impact and photographed by the accompanying fly-by spacecraft [4]. Using solar-electric propulsion or the solar sail, it would be possible to alter direct a payload into a retrograde solar orbit so that it would strike an offending NEO at a relative velocity of perhaps 60 km/s. Aim would have to be perfect since probe and NEO would be close for a small fraction of a second. Once again, certain NEO types might experience fragmentation rather than solar-orbital deflection.

Various researchers have proposed a number of solar-sail based NEO deflection schemes. At the same conference where the results of the 2007 NASA Marshall Space Flight Center study were reported, the Solar-Sail Gravity Tractor was discussed as well [5]. The Gravity Tractor functions by maintaining a small solar sail at a constant distance from the Earth-threatening

NEO. Application of Newton's Law of Universal Gravitation reveals that the NEO's solar trajectory will be slightly altered its attraction to the sail. Although the Gravity Tractor would work with any type of NEO, it might be necessary to maintain the sail's position for a period of decades.

If Earth-impact warning time is sufficiently long, a NEO could be encased in a solar sail [6]. This action, which would increase the NEO's reflectivity and surface area, could also convert an Earth-impact into a near miss if the impact warning time were measured in decades.

Another solar-sail based NEO deflection technique that could be applied if impact-warning time is measured in decades applied the Yarkovsky Effect [7]. Here, a flotilla of planar solar sails is used to either reflect additional sunlight on or shield portions of a NEOs surface from sunlight. The controlled asymmetry in incident and reradiated radiation pressure can, in principle, slowly alter the NEO's solar orbit.

The Solar Collector, SC, which was considered in the 2007 NASA Marshall study [3], also applies one or more solar sails to reflect light on a NEO's surface. In this case, the primary sail reflective element is parabolic or spherical and the sunbeam incident on the NEO is concentrated. As is discussed below, an energized jet will be raised from the surface of certain NEO types, which can divert a NEO's orbit. In many cases, the SC must remain on station for a period of years rather than decades, to convert a predicted Earth impact into a near miss.

Interestingly, the SC is one of the few proposed NEO deflection techniques that could be used to steer small NEO's into high-Earth orbit. As well as protecting the Earth, the SC could thus provide an opportunity to interests seeking to commercialize space by exploiting extraterrestrial resources.

THE SOLAR COLLECTOR (SC)

Figure 1 is a schematic representation of the SC's operation in NEO deflection. The SC is identical in configuration to a two-sail solar sail called the Solar Photon Thruster (SPT) [8-10]. A parabolic or spherical collector sail always faces the Sun. The concentrated beam from this sail falls upon the smaller, planar thruster sail . The thruster sail is adjusted so that the concentrated sunbeam is projected upon the NEO surface. This concentrated beam on the rotating NEO creates in volatile-rich NEOs an energized jet. By

Newton's Third Law, the reaction to this energized jet alters the NEO's solar trajectory.

Application of the solar photon thruster as an Earth-threatening NEO trajectory-deflector was suggested by Melsosh et al in 1994 . As well as presenting an initial theoretical basis for SC operation, they reported experimental results raising an energized jet from simulated NEO material using a pulsed laser in place of the concentrated sunbeam [11].

Melosh and his collaborators irradiated in these experiments a simulated NEO sample composed of basalt. The energy density of the laser-generated hotspot of the surface of the simulated NEO was about 10^9 watts/m^2 and the energized plume raised by the laser had a diameter of about 1 cm. The plume "exhaust velocity" was about 1 km/s [11].

This preliminary theoretical and experimental study was the basis for the SC model applied in the NASA Marshall 2007 NEO-deflection study [3]. In the preliminary NASA Marshall model, the energized-jet exhaust velocity was maintained at 1 km/s.

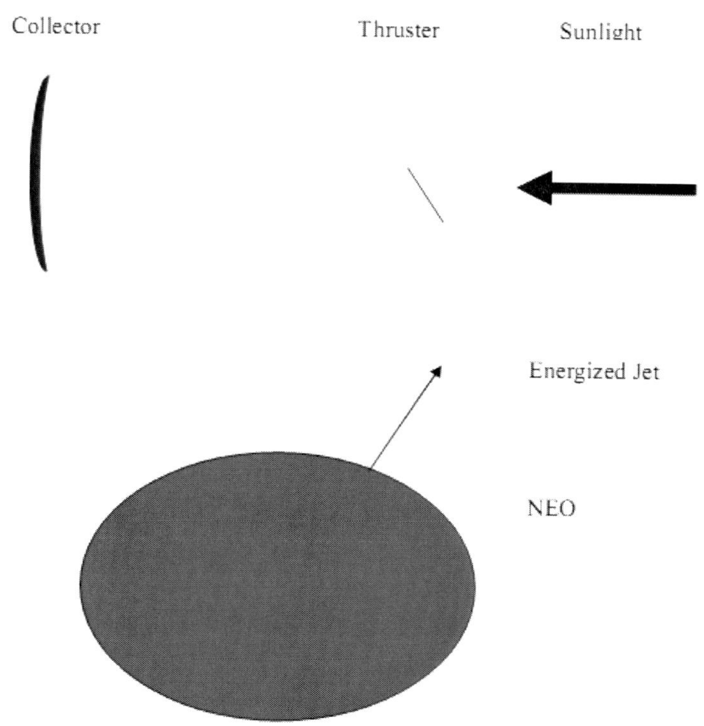

Figure 1. Schematic of Solar Collector Operation.

When this author began his consideration of the SC, he conducted additional analysis in an attempt to better understand the physics behind the energized jet exhaust velocity. Jet exhaust velocity was found to depend upon a number of parameters, including the electromagnetic (EM) penetration depth—the depth of NEO material to which solar radiation penetrates. A related literature search revealed that terrestrial soils have a typical EM penetration depth of 100 microns (0.1 mm). No information was discovered regarding EM penetration depth of extraterrestrial soils and surfaces [12,13].

Although it was demonstrated in this preliminary study that the major contribution to phase-change energy loss from the concentrated sunbeam for most materials was the heat of vaporization, heat loss by conduction was estimated. Another disadvantage of the preliminary model is that, from a dimensional point of view it was imperfect—it was implicitly assumed that NEO material would be exposed to the sunbeam for intervals of only one second [12,13].

The next section of this chapter presents an improved SC model. The new model is dimensionally correct, allowing consideration of arbitrary sunbeam exposure times for NEO surface material. Conduction of sunbeam energy is shown to be a major loss mechanism. Transmission measurements of an extraterrestrial sample are used to demonstrate that at least some NEOs will have an EM penetration depth similar to that of terrestrial soils.

MODELING THE SOLAR COLLECTOR AS A NEO DEFLECTOR

In modeling the SC, it is significant to remember that this device will concentrate sunlight like a magnifying glass rather than collimating it like a laser. Following the arguments of Bliss [14], the hotspot NEO area (A_{hs}) is related to the solar angular diameter α in radians and the separation between the Collector sail and the NEO surface in meters, a:

$$A_{hs} = \pi(a\alpha)^2 / 4 \tag{1}$$

For the case of a 1-km Collector-NEO separation and the Sun's angular size of 0.00928 radians, $A_{hs} = 67.6\ m^2$ and the hotspot radius is 4.64 m.

If is assumed that the angle between incident light and the thruster normal is 45°, a cosine factor reduces sunbeam intensity by 0.707. Contemporary sail material is assumed for both Thruster and Collector (reflectivity of sunlight = 0.9). Since contemporary sails can operate sunward of Mercury, only about 6% of the incident sunlight will be occulted by the thruster sail [15]. The solar power incident on the NEO (P_{in}) in the concentrated sunbeam is therefore 0.53X the solar power entering the solar collector. As discussed below, this may be conservative.

The next term defined is X_{neo}, the length that the hotspot center moves in time interval ΔT due to NEO rotation. Defining P_{neo} as the NEO rotational period in seconds and R_{neo} as the NEO's equivalent spherical radius,

$$X_{neo} = 2\pi R_{neo}(\Delta t)/P_{neo} \tag{2}$$

Figure 2 presents a schematic representation of the hotspot produced by the concentrated sunbeam on the NEO surface. It is assumed that the top of the hotspot, from which the energized jet escapes, is in the gaseous phase. The hotspot center is liquid and the hotspot bottom is solid.

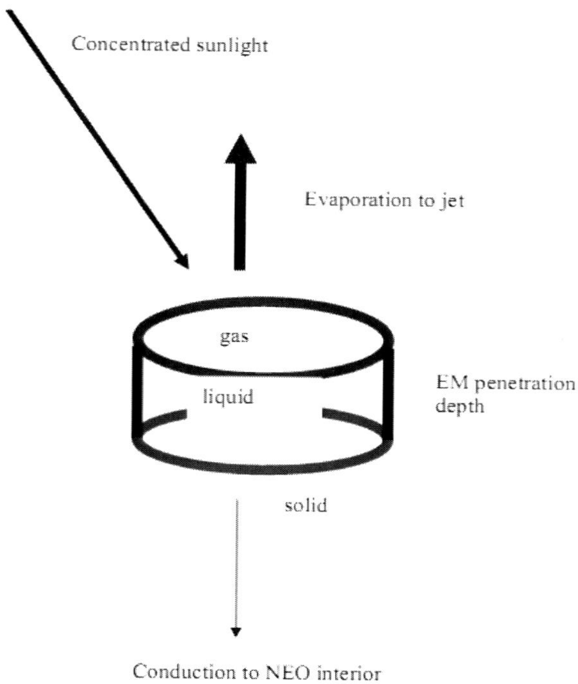

Figure 2. NEO Hotspot Schematic.

As previously discussed, the EM penetration depth approximates 100 microns. Because this is much smaller than the hotspot area in all practical cases, it is assumed that conduction heat loss to the NEO is through the bottom of the hotspot rather than through the sides.

Defining the concentrated sunbeam hotspot diameter as D, the EM penetration depth as t_{neo}, and the NEO material average density as ρ, the incremental NEO mass energized by concentrated sunlight during time interval Δt can be estimated:

$$\Delta M_{neo} \approx \frac{\pi}{4}\left(D + X_{neo}\right) D \rho t_{neo} \tag{3}$$

defining P_l as the incident hotspot power lost to phase change, conduction and radiation and Δt_h is the time interval that a typical hotspot point on the NEO is within the concentrated beam, the velocity of the energized jet ejected from the NEO can be expressed as:

$$V_{jet} = \left[\frac{2\left(P_{in} - P_l\right)}{\Delta M_{neo} / \Delta t_h}\right]^{1/2} \tag{4}$$

the time increment ΔT_h can be expressed as the ratio of hotspot diameter aα to tangential hotspot rotational velocity $2\pi R_{neo}/P_{neo}$. Therefore,

$$\Delta t_h = \frac{\alpha a P_{neo}}{2\pi R_{neo}} \tag{5}$$

in Ref.s. 11-13 and 15, it is assumed that the energized jet emerges from the NEO into a hemisphere. NEO acceleration due to this energized jet can be written:

$$ACC_{neo} = \frac{\Delta V_{neo}}{\Delta t_h} = \frac{0.5\left(\Delta M_{neo} / \Delta t_h\right)}{M_{neo}} V_{jet} \tag{6}$$

where the factor "0.5" is due to the perhaps questionable assumption of uniform jet emission into a hemisphere.

As discussed in Refs. 11-13 and 15, phase-change effects in the power-loss term P_l for most material are dominated by the heat of vaporization.

Phase-change heat loss is the product of NEO mass energized by the sunbeam per unit time and NEO material heat of vaporization. Radiation power-loss effects are only significant for high boiling-point materials and conductive losses will be almost entirely downward from the hotspot into the NEO. Conductive power loss can be estimated using the standard equation [16]:

$$P_{l,con} = -\frac{KA\Delta T}{t_{neo}} \tag{7}$$

where K is thermal conductivity, ΔT is temperature difference between hotspot top and bottom and A is sunspot area perpendicular to the sunbeam. For the case of a spherical NEO, $A = A_{hs}$.

DETERMINATION OF EM PENETRATION DEPTH FOR AN EXTRATERRESTRIAL SAMPLE

In the preparation of Refs. 12 and 13, a literature review revealed one study of the EM penetration depth of terrestrial soils. For the soils tested, EM penetration depth is about 100 microns (0.1 mm) [17]. Since the author serves as a Hayden Associate at the American Museum of Natural History (AMNH) in New York City as well as in his academic capacity, he arranged to borrow for analysis two meteorite samples from Dr. Denton Ebel, Curator of Meteorites at AMNH. These samples—a 30-micron thin section about 1.4 cm in diameter mounted on a glass slide and several grams of ground, simulated regolith, were from the Allende meteorite, a CV carbonaceous chondrite that impacted Mexico in 1969 [18].

To estimate EM penetration depth in these samples, apparatus was constructed and utilized to perform optical transmission measurements at two laser wavelengths: 532 and 650 nm. For both wavelengths and varied beam sizes, the transmission of the 30-micron thin section was 9-10%. For 532-nm light, average transmission of 30-micron and 50-micron regolith samples was respectively 2.35% and 0.59%. For 650-nm light, the same samples had a transmission of 5.78% and 0.79% [15,19].

It is hoped that other researchers will repeat these measurements at various wavelengths for numerous extra terrestrial samples. But on the basis of these experimental results, the EM penetration depth in this SC modeling study will be assumed to be 100 microns, identical to that of many terrestrial soils.

CONSTRAINTS ON SOLAR COLLECTOR DESIGN AND POSITION

As a two-sail solar sail in a dynamic deep-space environment, there are a number of design and position issues that the Solar Collector must contend with. The first of these is size.

In the initial paper by Melosh et al, it was assumed that a small SC of ~50-m radius would be adequate for the diversion of a small volatile-rich NEO [11]. But when solar-concentrator theory is applied to the problem, one learns that the SC-NEO separation for this device is a few hundred meters [12,13]. As will be justified below, a larger stand off distance may be necessary. So we assume here a SC Collector sail radius of about 140 meters.

If the Collector area is 13X the Thruster area [15], total sail area is 3.4 X 10^4 square meters. Assuming contemporary sail material, the collector and thruster sails have a specific gravity of 2 and a thickness of 7 microns. The area mass thickness of the sail material is therefore 0.014 kg/m^2 and the total sail mass is about 1000 kg. Assuming an additional 200 kg for structure and a 200-kg payload for control and communication, the total SC spacecraft has a mass of about 1400 kg.

The Collector sail of the SC will always face the Sun. Therefore, it is impossible to eliminate by solar sail technology alone solar-radiation-pressure force. This force, unless compensated for by external propulsion, will tend to push the SC off station. Applying solar-sail theory and assuming a constant Collector (and Thruster) reflectivity of 0.9 and a 1-AU position for the SC and NEO with a solar constant of 1,366 watts/m^2, maximum solar radiation pressure on the Collector sail is calculated to be about 4 X 10^{-4} m/s^2 (or 4 X 10^{-5}g). [20,21]. If a solar-electric thruster with an exhaust velocity of 30 km/s is attached to the SC, only about 10^{-5} kg/s of ion fuel must be expelled each second to maintain the SC on station.

Only a few hundred kilograms of ion fuel are required to maintain the SC on station for a multi-year time interval. This may be an over estimate since solar radiation pressure acceleration on the Thruster will tend to partially compensate for solar radiation pressure on the Collector.

In the following sections of this chapter, Collector-NEO separation is maintained at about 1 km. This is because, in a recent model study, Gong et al has considered lifetime-limits on an SC immersed in the energized plume raised by the concentrated sunbeam on the NEO surface. Although Gong et al conclude that a ~1km SC-NEO separation is adequate, additional study is required [22]. For example, is the Melosh et al assumption of a hemispheric energized jet reasonable [11]? Might the jet instead be Gaussian? And how is jet configuration affected by NEO rotation?

ASSUMED NEO PROPERTIES

Human knowledge of the NEO population is currently fragmentary at best. We know from museum collections that meteorites, which originate from the NEO population, may be iron rich, stony (silicate rich) or volatile rich carbonaceous chrondites. Iron-rich meteorites are likely over represented because silicates erode more rapidly on the terrestrial surface and many carbonaceous chrondites vaporize like the Tunguska object before reaching the Earth's surface.

According to Rabinowitz et al [23], fully 50% of observed NEOs have spectral similarity to volatile-rich objects. Remo concludes that volatile-rich NEOs are not uncommon [24]. This is a good thing from the point of view of the Solar Collector. In Ref. 15, a 50-m solar collector a few hundred meters from a NEO is considered for diverting water-ice rich, iron-rich and silicate-rich NEOs. Because of the high melting point of iron, the SC will be ineffective in diverting such an object. A considerably larger SC will be required to divert a silicate-rich NEO. The SC is most effective for diversion of ice balls.

We here assume that the NEOs to be diverted by the SC are similar to extinct comet nuclei. These objects have caused ~20-30% of impacts on Earth during our planet's history [25]. For simplicity, we assume here that the NEOs in question are composed of ice layers over porous rock. The specific gravity of the structure is assumed to be 1.

From Ref. 25, most comet nuclei rotate with a period of days. However, close encounters with terrestrial planets may have affected NEO rotation rates. About 10% of observed NEOs rotate every ~28 hours, 23% have an average rotation rate of about 11.3 hours [26]. We select a rotation rate in this analysis of 50,000 seconds.

In this chapter, we consider two NEOs—one for diversion and a smaller one for exploitation. The larger one has an equivalent spherical radius of 200 m and a specific gravity of 1. The mass of this object is approximately 3×10^{10} kg. The smaller object has an equivalent spherical radius of 20 m, a specific gravity of 1 and a mass of about 3×10^{7} kg.

MINIMUM SOLAR COLLECTOR SIZE

As demonstrated in Ref. 15 and in the case studies below, the major input power loss for sunlight concentrated by the SC on an ice-rich NEO is conduction. We therefore require that $P_{in} > P_{l,con}$. The input power to the hotspot can be defined as:

$$P_{in} = 1366 f \pi R_c^2 \quad \text{watts/m}^2, \tag{8}$$

where f is the fraction of incident sunlight on the Collector sail concentrated in the hotspot and R_c is the Collector radius (which is here assumed to be disc shaped from the viewpoint of incident sunlight).

We assume that the top of the hotspot has a temperature of 373 K, the bottom of the hotspot has a temperature of 245 K and the EM penetration depth of the NEO material is 10^{-4} m. The thermal conductivity of the hotspot is assumed equal to that of liquid water, 0.58 W/m-K [15]. Substitution in Eq. (7) yields

$$\frac{fR_c^2}{A_{hs}} > 173 \tag{9}$$

If the NEO is at 1 AU and the SC-NEO separation is 1 km, the hotspot area is 67.6 m^2 as discussed above. Therefore,

$$fR_c^2 > 11700 \tag{10}$$

The next parameter to be investigated is f, the fraction of sunlight incident on the large Collector sail that is concentrated in the NEO hotspot. This factor is the product of Collector sail reflectivity, Thruster sail reflectivity, the light fraction not occulted by the thruster and the cosine of the angle between the light reflected from the Collector sail and the normal to the Thruster sail.

We assume that the reflectivity of both sails to sunlight is 0.9 and that 6% of the incident sunlight is occulted by the Thruster sail [15]. The cosine factor is ~0.71 for a ~45-degree angle between Collector-reflected light and Thruster normal, as shown in Figure 1. But one can imagine an SC positioned sunward of the NEO with Cassegrain optics. In this case, the cosine factor may be as high as ~0.9. Thus, f can vary between 0.54 and 0.69. The radius of an SC positioned ~1 km from the NEO can realistically be expected to be in the range 131-147 m.

SCENARIO 1: DIVERTING AN EARTH-THREATENING NEO

We consider in this section the diversion of a 200 m radius spherical water-ice-rich NEO with a specific gravity of 1, a mass of 3×10^{10} kg, and a rotation period of 50,000 seconds. To successfully convert an Earth-impact into a near miss, the SC must alter the NEO's solar orbit by 6,000 km, one Earth radius.

The SC's Collector sail is positioned 1 km from the NEO surface. The SC has a radius of 149 m and 54% of the sunlight striking the SC collector sail is concentrated into the hotspot on the NEO surface. A total of 9.53×10^7 watts of sunlight is incident upon the Collector sail. Of this, 5.15×10^7 watts is concentrated in the hotspot. The hotspot area is 67.6 m^2; the hotspot radius is 4.64 m. Substitution in Eq. (5) reveals that $\Delta t_h = 369$ seconds. From Eqs. (2) and (3), $\Delta M_{neo} = 13.5$ kg. The total energy in the hotspot is approximately 1.9×10^{10} Joules.

The first power-loss term is the amount of energy required to raise the temperature of 13.5 kg of water ice from 245 to 273 K. Since the specific heat of water ice is 2.1 kJ/kg-K [15], 794 kJ of concentrated solar energy must be applied to this process.

Since the heat of fusion of water is 333 kJ/kg [15], an additional 4496 kJ is required to melt the ice. The specific heat of liquid water is 4.2 kJ/kg [15]. Therefore, an additional 5670 kJ is required to raise the water temperature to 373 K.

The heat of vaporization of water is 2256 kJ/kg. To evaporate the 13.5 kg of liquid water into the energized jet requires 30,456 kJ.

Before considering conduction, the energy loss to evaporate water into the jet is 4.14×10^7 J. This is about 0.0017 of the concentrated solar energy on the hotspot. The available hotspot power excluding conduction is 5.14×10^7 J.

Conduction power loss ($P_{l,con}$) is calculated using Eq. (7). Substituting in this equation, $P_{l, con}$ is approximately equal to 5.02×10^7 Watts. The net solar power to the jet ($P_{in} - P_l$) is therefore approximately equal to 1.2×10^6 Watts.

Equation (4) can now be used to estimate the velocity of the energized jet. Jet velocity is about 8 km/s. As discussed in Ref. 15, this is close to the initial velocity of comet tails [27, 28].

For the case of an energized jet emitted into a hemisphere, Eq. (6) can next be applied to estimate the jet-induced acceleration of the NEO. This acceleration is approximately 5×10^{-9} m/s^2. In one year, the SC-energized jet can alter the NEO's solar-orbit velocity by about 0.15 m/s. If the SC can be maintained on station for a few years, an Earth-impact can be converted into a near miss.

During every second, about 0.04 kg of NEO material enters the jet. In every year, this is equal to an ejected mass of about 10^6 kg, less than one part in ten-thousand of the NEO mass prior to SC operation. Next we evaluate utilization of the Solar Collector to exploit the resources of a smaller NEO by applying a velocity increment sufficient to move it into a high Earth orbit.

Scenario 2: Steering a Small NEO into High Earth Orbit

Consider here the 20-m diameter ice ball with a specific gravity of 1 and a mass of 3×10^7 kg. This object also has an assumed rotation rate of 50,000 seconds. Scaling from the previous scenario, we see that the NEO acceleration imparted by the SC-excited jet is 5×10^{-6} m/s^2. In one year, a velocity increment of about 160 m/s will be produced.

If a NEO solar-orbital-velocity change of 2 km/s is required to steer the NEO into high Earth orbit, the Solar Collector must remain on station for 12.5 years. During this period of time, about 50% of the NEO's mass is exhausted in the energized jet.

About 1.5×10^7 kg of NEO material will be available to astronaut and robot miners in high Earth orbit. This could be used for habitat oxygen and water, rocket fuel and galactic cosmic ray shielding. Whether such NEO-resource mining is more economical than direct launch from Earth surface cannot be determined in this study.

Conclusion

The model described here is somewhat congruent with experimental evidence. In 2002, Gregory Benford and James Benford described "desorption." An intense electromagnetic beam can "boil off" volatile materials within the coatings of solar sails at high velocity, thereby boosting sail performance [29]. This process can be thought of as an analog to the Solar Collector's application in raising an energized jet of volatile material from a NEO's surface.

No model is a perfect simulation of reality. For instance, it is very doubtful that pure-water-ice NEOs will be found. Probably, NEOs that are extinct comet nuclei will contain layers of water ice infused with suspended regolith particles. This will probably reduce the plume jet velocity and extend the Solar Collector's on-station time required to divert or collect a selected NEO.

Also, one wonders about Solar Collector lifetime estimates. For example, consider the case of a Cassegrain-optics Solar Collector stationed closer to the Sun than the NEO. Plume particles evaporated from the NEO in the direction of the Solar Collector will encounter solar photons and photons in the concentrated beam. Radiation pressure from these photons will tend to decelerate evaporated jet particles directed towards the Solar Collector, thereby reducing Solar Collector optics erosion and increasing solar Collector life expectancy.

In all likelihood, no model will satisfactorily resolve all the issues related to the employment of the Solar Collector for NEO diversion or retrieval. As with all the suggested methods of Earth defense, the Solar Collector must be tested on a NEO. Since human expeditions may visit nearby NEOs within a decade or so, it is hoped that NEO-diversion techniques can be tested by astronaut crews in the not-remote future.

ACKNOWLEDGMENTS

The author appreciates discussions with and the assistance of R. Adams, D. Ebel, L. Johnson, R. Kezerashvili, and L. Leng.

REFERENCES

[1] Yeomans, D. and Chamberlin, A. (2011). *Near-Earth Asteroid discovery statistics.* http://neo.jpl.nasa.gov.stats.

[2] Simonenko, V. A., Nogin, V. N., Petrov, D. V., Shubin, O. N., and Solem, J. C. (1994). Defending the Earth against impacts from large

comets and asteroids. *In Hazards Due to Comets and Asteroids*; Gehrels, T.; Ed. University of Arizona Press, Tuscon, AZ; pp. 929-953.

[3] Adams, R. B., Campbell, J. W., Hopkins, R., Smith, S., Arnold, W., Baysinger, M., Crane, T., Capizzo, P., Sutherin, S., Dankovich. J., Woodcock, G., Edlin, G., Rushing, J., Fabisiniki, L., Jones. D., McKamey, S., Thomas, S., Maccone, C., Matloff, G., and Remo, J. (2007*). Continuing efforts at MSFC to develop mitigation technologies for near-Earth objects.* Presented at the AIAA/Aerospace Corporation 2007 Planetary Defense Conference, George Washington University, Washington, D.C., March 5-8, 2007.

[4] Simpson, C. (2005). Deep Impact's smash and grab. *Spaceflight,* 47, 344-347.

[5] Wie, B. (2007). Hovering control of a solar sail for asteroid deflection (2007). Presented at the AIAA/Aerospace Corporation 2007 Planetary Defense Conference, George Washington University, Washington, D.C., March 5-8, 2007. Also presented as Paper AAS07-145 at the 2007 AAS/AIAA Spaceflight Mechanics Meeting, Sedona, AZ, January 28-February 1, 2007.

[6] Matloff, G. L. (1999). Applying International Space Station and solar sail technology to explore/divert small, dark NEOs. *Acta Astronautica,* 44, 151-158.

[7] Prado, J.-Y., Perret, A., and Boisard, O. (2010). Deflecting Apophis with a flotilla of solar shields. Presented at the Second International Symposium on Solar Sailing, New York City College of Technology, Brooklyn, NY, July 20-22, 2010. *Published in Proceedings of Second International Symposium on Solar Sailing;* Kezerashvili, R., Ed.; pp. 103-109.

[8] Malanin, V., and Repyakh, A. (1974). On the motion of a craft with two solar sails. *Problemy Mekhaniki Upravlyayemogo Dvizheniya*, No. 5, 99-108.

[9] Forward, R. (1990). Solar photon thruster. *Journal of Spacecraft and Rockets*, 27, 411-416.

[10] Dachwald, B. and Wurm, P. (2010). Mission analysis and performance comparison for an advanced solar photon thruster. Presented at the Second International Symposium on Solar Sailing, New York City College of Technology, Brooklyn, NY, July 20-22, 2010. *Published in Proceedings of Second International Symposium on Solar Sailing;* Kezerashvili, R., Ed.; pp. 147-156.

[11] Melosh, H. J., Nemchinov, I. V., and Zetzer, Yu. I. (1994). Non-nuclear strategies for deflecting comets and asteroids. In *Hazards Due to Comets and Asteroids;* Gehrels, T.; Ed. University of Arizona Press, Tuscon, AZ; pp. 1111-1132.
[12] Matloff, G. L. (2008). The solar collector and near-Earth object deflection. *Acta Astronautica*, 62, 334-337.
[13] Matloff, G. L. (2009). The solar collector and near-Earth object diversion. *In Solar Collectrors: Energy Conservation, Design, and Application*; Killian, A. V., Ed. Nova Science Publishers, NY, pp. 221-227.
[14] Bliss, R. (1957). Designing solar furnaces for specific performance. *Solar Energy*, 1, 55-62.
[15] Matloff, G. L. (2011). Deflecting Earth-threatening asteroids using the solar collector: An improved model. Presented at Missions to the Outer Solar System and Beyond: *Seventh IAA Symposium on Realistic Near-Term Advanced Scientific Space Missions*, Aosta, Italy, July 11-14, 2011.
[16] Ohanian, H. C. (1985). *Physics,* 2nd. ed. Norton, NY; pp 524-527.
[17] Ciani, A., Goss, K.-U.. Schwartzenbach, R. P. (2005). Light penetration in soil and particulate minerals. *European Journal of Soil Science*, 56, 561-574.
[18] Bevan, A. and De Laeter, J. (2002). *Meteorites: A Journey Through Space and Time.* Smithsonian Institute, Washington, D.C., pp. 84-99.
[19] Matloff, G. L., Leng, L., and Le, T. (2010). Optical transmission of an Allende meteorite thin section and simulated regolith. Presented at 73rd meeting Meteoritical Society, American Museum of Natural History and Park Central Hotel, New York, July 26-30, 2010. *Abstract published in Meteorites and Planetary Science Supplement*, 45, A127 (July, 2010).
[20] Matloff, G. L. (2005). *Deep-Space Probes*, 2nd ed. Springer-Praxis, Chichester, UK, pp. 49-51.
[21] Vulpetti, G., Johnson, L., and Matloff, G. L. (2008). *Solar Sails: A Novel Approach to Interplanetary Travel,* Springer-Copernicus, NY and Springer-Praxis, Chichester, UK, pp. 171-184.
[22] Gong, S.-P., Li, J.-F., and Gao, Y.-F. (2011). Dynamics and control of a solar collector system for near Earth objects deflection. *Research in Astronomy and Astrophysics*, 11, 205-224.
[23] Rabinowitz, D., Bowell. E., Shoemaker, E., and Muinonen, K.(1994). The population of Earth-crossing asteroids. In *Hazards Due to Comets*

and Asteroids; Gehrels, T.; Ed. University of Arizona Press, Tuscon, AZ; pp. 285-312.

[24] Remo, J. L. (1994). Classifying and modeling NEO material properties and interactions. In *Hazards Due to Comets and Asteroids*; Gehrels, T.; Ed. University of Arizona Press, Tuscon, AZ; pp. 551-596.

[25] Rahe, J., Vanysek, V. and Weissman, P. R. (1994). Properties of cometary nuclei. In *Hazards Due to Comets and Asteroids*; Gehrels, T.; Ed. University of Arizona Press, Tuscon, AZ; pp. 597-634.

[26] Scheeres, D. J., Marzari, F., Rossi, A. (2004). Evolution of NEO rotation rates due to close encounters with Earth and Venus. *Icarus,* 170, 312-323.

[27] Vanysek, V., Grygar, J., and Sekanina, Z. (1959). The initial velocity in the tail of Comet 1956B. *Bulletin of the Astronomical Institute of Czechslovakia,* 10, 116-120.

[28] Brandt, J. C. (1982) Comets. In *The New Solar System*, 2nd ed.; Beattly, J. K., O'Leary, B., and Chaikin, A. ; Eds. Cambridge U. P., NY.

[29] Benford, G. and Benford, J. (2002). "Desorption-assisted Sun Diver missions." Presented at STAIF-2002. *Published in Proceedings Space Technology and Applications International Forum*; Ed. M. S. El-Genk, American Institute of Physics.

In: Meteorites and Asteroids
Editors: A. Dementieva et al.
ISBN: 978-1-61942-115-8

Chapter 4

RADAR CHARACTERISTICS OF ASTEROID 33342 (1998 WT24)

Yu. A. Gavrik[1*], A. L. Gavrik[1*] and V. A. Sorokin[2]
[1] Kotel'nikov Institute of Radio Engineering and Electronics of RAS, Fryazino, Russia
[2] Joint Stock Company «Distant Radiocommunication Scientific Research Institute» Moscow, Russia

ABSTRACT

In 2001 during the closest approach to the Earth asteroid 1998 WT24 was observed by Evpatoria–Medicina bistatic radar system. We present obtained radar characteristics of this object based on this observation. In this chapter variations in the power spectra of echo signals are analyzed for two orthogonal polarizations of the echo signal. The analysis of the asteroid scattering characteristics for different polarizations indicates stable polarization ratio for the asteroid aspect angles comprising almost the entire asteroid surface. The result of simulation has shown that the scattering law degree of the asteroid roughness is ~1.8; however, the surface is nonuniformly rough. The obtained data also were used for estimation of physical properties of the asteroid showing an average diameter of 405±10 m and large-scale features of the relief on the asteroid surface. The result of measurements of the radar albedo and the surface

[*] E-mail: alg248@ire216.msk.su

roughness allow us to present estimations about the properties of the surface material of the asteroid.

INTRODUCTION

Asteroid (33342) 1998 WT24 was discovered on November 25, 1998, by LINEAR about four months after its close encounter with the planet Mercury. The 2001 approach to 0.0125 AU provided an excellent opportunity for ground based investigations of the asteroid by radar and optical systems. During its close Earth approach in December 2001, 1998 WT24 was observed extensively using thermal infrared radiometry [1], optical photometry and spectroscopy [2, 3, 4] and radar sounding [5-7].

In this chapter we will analyze result of radar observations organized by M. Di Martino, S. Ostro, A. Zaitsev [6,7]. This radar sounding of the asteroid was performed by Evpatoria–Medicina bistatic radar system on December 16 and 17, 2001 with a monochromatic wave ($\lambda \approx 6$cm). The radar system consisted of the 70-m parabolic antenna in Evpatoria Space Center (Ukraine) which was used as a transmitter radiating an unmodulated, continuous-wave, clockwise circularly polarized signal with frequency 5010 MHz and power 90 kW, and a 32-m parabolic antenna with a low-noise two-channel polarization receiver in Medicina Radio Astronomical Station (Italy). The characteristic of the system and experiment details are described in [6,7].

RADAR DATA ANALYSIS

During the radar observation of asteroid 1998 WT24 the transmitted frequency was shifted on a prior predicted Doppler shift in order to avoid spectral smearing of radar echo. Received echo was split into clockwise and counterclockwise polarizations, heterodyned, filtered and then recorded with the sampling rate either 10 kHz or 50 kHz [7].

Initial radar data analysis showed that despite of the predicted Doppler shift compensation, the difference between the predicted and measured variations in the Doppler frequency shift in power spectra could reach 1.5 Hz [8]. Performed correction of this Doppler shift allowed us to increase a signal to noise ratio by extending of the spectrum accumulation time in our processing technique. As a result, spectrum widening caused by asteroid rotation was determined more accurately and the influence of noise became minimal under the conditions of the considered experiment. Further analysis revealed that all 6 experiment sessions had substantial variations in the noise level of the receiving equipment [7]. These variations were neutralized by means of equalization of the fluctuations variance of the echo-signal amplitude. Also, it was found that in sessions with a sampling rate of 50 kHz, the signal- to-noise ratio was almost 2 times higher than that in sessions with a sampling rate of 10 kHz due to some issues with the equipment [7]. Thus, in the subsequent detailed analysis, we mainly used sessions with a sampling rate of 50 kHz while only reliable data was taken from other sessions.

A special technique was used to achieve high signal-to-noise ratio in order to get accurate scattering properties of the asteroid surface from the power spectra. Decreasing the noise and increasing accuracy of the measured power spectra parameters were attained by heterodyning, filtering, and reducing the bandwidth of the detected echo signals to the bandwidth 156 Hz. Then, processed complex echo signals were separated into overlapping segments, and were multiplied by a window function to suppress sidelobes influence, and were supplemented with zero samples in order to increase resolution of the fast Fourier transformation [9]. For each segment, the echo-signal power spectra were centered at a frequency 0 Hz with bandwidth of ±78 Hz and obtained resolution of the power spectra were 0.15 Hz [8]. Finally, these spectra were accumulated within a time interval of 3 min.

Obtained power spectra are shown on Fig 1 and main parameters of these spectra are listed in Table 1 [8]. The noise characteristics are described by parameter ΔP_N, which is equal to the value of noise fluctuations relative to the average level normalized to unity. Parameter W_Σ is the normalized value of the integrated energy of the echo-signal power spectrum, parameter W_{max} corresponds to the energy of the maximum spectral component, and B_- and B_+ are the spectrum boundaries in the negative and positive frequency regions. Angle zero corresponds to asteroid axis orientation at the beginning of the experiment and asteroid the rotation period is 3.697 ± 0.001 h based on optical observations [3].

As seen from the data in sessions 2–4 (see Table 1 and Figure 1), integrated energy of power spectra vary with the asteroid aspect angle. The maximum value was observed in session 3b and the minimum value of the reflected energy was observed in session 2b for both polarizations.

The difference between the energies is a factor of 1.5 for the OC polarization and a factor of 1.2 for the SC polarization. High energy variations at the OC polarization point to the existence of regions of mirror reflection.

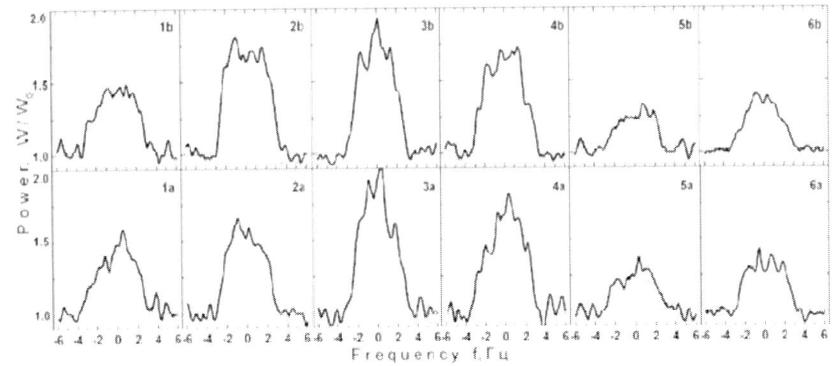

Figure 1. Power Spectra for OC polarization (solid line) and SC polarization (dotted line).

In all sessions, the ratio of the integral signal energies at two polarizations varies from ~0.9 to ~1.1 and corresponds to the values of the measured polarization ratio (~1 .0) [8]. Variations in W_{max} do not exceed a factor of 1.5 and could be related to the structural features of the reflecting surface.

Table 1. Main Parameters of the power spectra

Session	Angle	Polarization	ΔP_N	W_Σ	W_{max}	B_-	B_+
1	10-50	SC	0.0407	12.83	0.442	−3.5	3.2
		OC	0.0381	14.66	0.488	−3.4	3.8
	50-89	SC	0.0507	12.45	0.490	−3.4	3.6
		OC	0.0546	12.08	0.403	−3.4	3.6
2	143-190	SC	0.0381	18.79	0.769	−3.3	3.0
		OC	0.0397	16.49	0.658	−3.1	3.3
	190-240	SC	0.0373	22.08	0.816	−3.3	3.2
		OC	0.0374	23.76	0.799	−3.0	3.6
3	280-310	SC	0.0446	22.61	0.912	−3.0	3.3
		OC	0.0374	23.08	1.007	−3.3	3.3
	310-337	SC	0.0402	20.33	0.891	−3.0	3.2
		OC	0.0434	21.42	0.930	−3.3	3.2

Session	Angle	Polarization	ΔP_N	W_Σ	W_{max}	B_-	B_+
4	18-40	SC	0.0477	22.08	0.836	−3.6	3.3
		OC	0.0486	20.19	0.817	−3.6	3.3
	40-58	SC	0.0476	21.22	0.829	−3.2	3.3
		OC	0.0475	20.61	0.718	−3.3	3.4
5	109-150	SC	0.0333	6.91	0.309	−3.2	2.8
		OC	0.0344	7.92	0.371	−3.1	3.2
	150-190	SC	0.0313	7.65	0.302	−3.0	3.0
		OC	0.0316	7.74	0.329	−3.2	3.0
6	201-250	SC	0.0322	10.89	0.450	−3.0	3.2
		OC	0.0326	10.09	0.412	−3.0	3.2
	250-298	SC	0.0335	9.37	0.435	−3.0	3.0
		OC	0.0327	10.1	0.401	-3.3	3.3

Variations in the spectrum boundaries that are caused by variations in the asteroid aspect angle do not exceed 14%, thus, the asteroid body do not have a strong asymmetry.

CHARACTERISTICS OF THE ASTEROID

Using spectrum characteristics from Table 1, the asteroid diameter can be determined from the relationship [10,11]:

$$B = \frac{4\pi D}{\lambda T}\cos\delta ,$$

where B is the width of the echo-signal power spectrum, D is the asteroid diameter, δ – angle between the spin axis and perpendicular to the Earth-asteroid line, and T is the asteroid rotation period. Having T as 3.6977 and predicted pole orientation and axis angle, the determined values of the maximum (D_{max}= 430±10 m), minimum (D_{min}= 380±10 m), and average (D_{avg}= 405±10 m) diameters of the asteroid almost coincide for both polarizations. This result means that the equatorial silhouette of the asteroid slightly deviates from a circle; however, this deviation does not exceed 14%.

These measurements can be used to estimate the value of optical albedo p_v as a function of asteroid diameter D and absolute stellar magnitude H [12]:

$$\log p_V = 6.247 - 2\log D - 0.4H$$

Taking into account the result from [13], where H = 18.54, p_v will be 0.41±0.02. However, using the data of other measurements (H = 17.90 [14]), we obtain a substantially different value of the optical albedo p_v=0.74±0.04. The cause of this difference from the data of one of the measurements is not yet revealed.

Another important characteristic of the asteroid is radar reflectivity which depends on its geometry, the structure, and the electric properties of the surface material of the asteroid. This parameter is characterized by the radar cross section and the radar albedo for the OC channel. The radar cross section is the cross section of a perfectly conducting metal sphere which, being placed instead of the asteroid, creates the same energy of the backscattered radar signal as that of the object. Radar cross section σ is determined from the relationship [15]:

$$\sigma = 4\pi\lambda^2 \frac{(R_t R_r)^2 k T_n \Delta f}{A_t A_r P_{rad}} \frac{s}{n}$$

where R_t and R_r are the distances from the transmitting and receiving antennas to the asteroid; A_t and A_r are the areas of the transmitting and receiving antennas; P_{rad} is the radiated power; T_n is the receiver's noise temperature; k is the Boltzmann constant; Δf is the frequency resolution of the spectra; and s/n is the signal-to-noise ratio. The surface roughness is characterized by polarization ratio μ, radar albedo n_{OC} and total albedo n_Σ [16]

$$\mu = \frac{\sigma_{SC}}{\sigma_{OC}} \qquad n_{OC} = \frac{\sigma_{OC}}{A} \qquad n_\Sigma = n_{OC} + n_{SC}$$

where σ_{SC} and σ_{SC} are cross section for two orthogonal polarization, A is a cross-sectional area.

The values of the radar cross section 0.021±0.002 km^2, the radar albedo $n_{OC} \approx 0.15$, and the total albedo $n_\Sigma \approx 0.30$ demonstrate high stability at different aspect angles of the asteroid. The reflectivity of asteroid 1998 WT24 was estimated with the use of only the data obtained in sessions 2, 3, and 4, in which the systematic errors caused by recording technique were minimal. The polarization ratio likewise demonstrates stability at the asteroid aspect angles comprising almost the entire asteroid surface; the value of this parameter is $\mu \approx 1.05\pm0.1$. It should be noted that, in sessions 1, 5, and 6, the value of parameter μ does not differ from the values measured in sessions 2, 3, and 4 while the radar cross section measured in sessions 1, 5, and 6 is about 2 times smaller due to the influence of systematic errors.

The next important characteristic of a surface is roughness. In order to estimate the surface roughness of asteroid 1998 WT24, we considered the result of numerical simulation described in [17, 18]. As a model we took a rough ellipsoid revolving with a period of 3.6977 h. The principal axes of this ellipsoid correspond to our estimates of the minimum and maximum dimensions of asteroid 1998 WT24. The diffuse scattering is usually described by the law $d\sigma/dA \sim \cos^n\theta$, where σ is the radar cross section, A is the surface area, and θ is the angle between the direction of radio waves and the normal to the surface. For uniform scattering, n = 1, and for Lambert scattering, n = 2. For most asteroids and planets, the value of parameter n ranges from 1.0 to 1.5 [19]. The result of the computation is shown on Figure 2 as a thin dotted line. The solid curves represent the experimental echo-signal power spectra and thin dotted curves show the result of simulation performed for n = 1.8 and the (a, c) OC and (b, d) SC polarizations at the beginning and the end of sessions 2, 3, and 4. Frequency f = 0 Hz corresponds to scattering from the asteroid's center of gravity. The echo-signal powers are normalized to the average noise level.

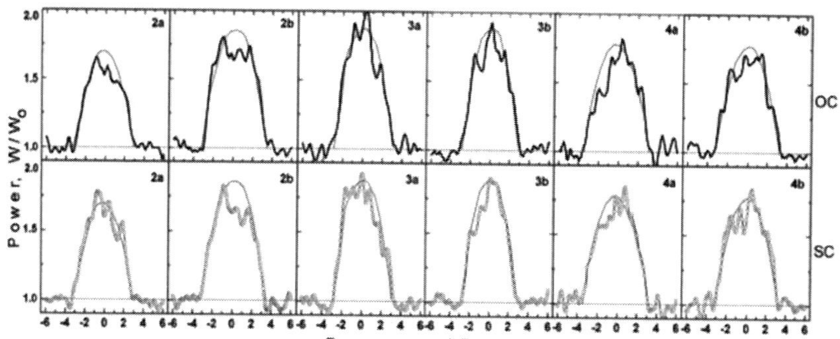

Figure 2. Comparison between (solid curves) the experimental echo-signal power spectra and (dashed curves) the result of simulation [8].

An optimal match between the model and asteroid 1998 WT24 is obtained if the axes of the ellipsoid are 430, 380 and 405 m and the scattering law is n=1.8±0.1. It is easy to see on Figure 2 a low half of the modeled and real spectra are the same while top halves occasionally have differences. Moreover the differences are observed in symmetrical and asymmetrical spectra. The main reason of such deviations is the surface of the asteroid is not convex and uniformly rough; it has large-scale features on the surface.

Analysis of the surface roughness reveals the coincidence of the boundaries and the shapes of the power spectra at different polarizations indicating possible diffuse scattering from the surface; however, the observed variations in the details of the power spectra corresponding to the different asteroid aspect angles do not confirm this statement.

For example, some spectra have distribution of the echo-signal energy as symmetric relative to the center of a spectrum while others have the distribution as asymmetric. Moreover, at certain angles, the power spectra contain local maxima of the signal energy, while the result of calculation using the rough ellipsoid demonstrates substantial discrepancies with the experimental data.

All these facts indicate the presence of large-scale features of the relief on the asteroid surface whose scattering properties depend on the asteroid's aspect angle.

Images of asteroid 1998 WT24 obtained in [5] clearly show the object structure having a complex structure of the asteroid surface. The result of our simulation confirms that the asteroid surface has complex configuration and is strongly rough, the scattering of its surface being described by a law close to the Lambert law.

RESULTS COMPARISON

Table 2 provides comparison of characteristics of (33342) 1998 WT24 obtained by different methods of observation [5, 6, 1, 2].

Presented in this chapter estimated values of the asteroid diameter agree with the result of radar sounding at a wavelength of 3.5 cm [5], the result of observations performed at a German observatory [1], and the data of infrared thermal surveying [2,3,4]. However, different approach of radar data processing gave the frequency resolution as 0.15 Hz, so the uncertainties are smaller. The polarization ratio measured in all sessions is stable for whole asteroid's aspect angles. Similar values of the polarization ratio were obtained at a wavelength of 3.5 cm during radar sessions performed with the use of the Goldstone (United States) [5] and Medicina (Italy) [6] radars.

However, we should mention that, in these experiments, the radar cross section was found to be almost two times larger than the radar cross section obtained in our experiment.

Table 2. Comparison of characteristics of (33342) 1998 WT24

	Present work	Busch et al. [5]	Martino et al. [6]	Harris et al. [1]	Kiselev et al.[2]
Maximum diameter (km)	0.430±0.01	0.470±0.04	-	-	0.420
Minimum diameter (km)	0.380±0.01	0.400±0.04	-	-	0.330
Effective diameter (km)	0.405±0.01	0.415±0.04	~0.410	0.35±0.04	[1] 0.40±0.13 [2] 0.34±0.10
Optical albedo	0.41±0.02 (H=18.54); 0.74±0.04 (H=17.90)	0.34±0.20 (H=18.69±0.3)	-	0.56±0.2	[1] 0.43±0.15 [2] 0.62±0.17
Mean radar cross-section (km^2)	0.021±0.002	0.027±0.003	~0.02	-	-
Radar albedo	0.15	-	~0.1	-	-
Total (SC+OC) radar albedo	0.30	0.42±0.04	-	-	-
SC/OC	1.05±0.1	0.97±0.10	~1	-	-
Density (g/cm^3)	~2.3	~3	-	-	-
Mass	$\sim 7.8 \cdot 10^{10}$	-	-	-	-
Radar scattering law index	1.8±0.1	OC:1.62±0.1 SC:1.68±0.1	-	-	-

[1] Lupishko and Mohamed, 1996, calibration;

[2] Cellino et al.,1999, calibration;

The same discrepancy of the radar cross section was obtained during radar observations of 6489 Golevka when a similar bistatic radar system was used [34]. The cause of the discrepancies is still unclear.

Asteroid 1998 WT 24 are thought to be E-class asteroid having enstatite achondrite composition [20], however a polarization ratio is atypical. Only a few asteroids: 2005 WC1, 17511 (1992 QN), 2000 EE104, 2102 Adonis, 2002 VE68, and 3103 Eger have similar polarization ratios [21]. At the same time, the same anomalously high values of the polarization ratio were observed during the radar sensing of the Jovian ice satellites (Europe, Ganymede, and Callisto) [22] and the poles of Mars [23] and Mercury [24], as well as during the sensing of a basalt lava stream on the Earth [25], the Alpha region on Venus [26], the largest volcanic plateau (Tharsis) on Mars [27], and a crater on the Moon [28]. Such high level of reflection in the SC channel can be caused by both multiple scattering and single scattering from a surface containing many scattering centers with curvature radius comparable with the wavelength. Even though the properties of asteroid 1998 WT24 are comparable with the properties of mentioned the Jovian satellite (μ = 1.1–1.3, n_{OC} ~ 0.3), we cannot be certain that this asteroid consists of ice, as Callisto does.

In contrast, the radar characteristics of asteroid 1998 WT24 are strongly different from the characteristics of the most of the observed asteroids while neither the orbit of this asteroid is comet like nor the radar characteristics of the asteroid agree with the characteristics of the already studied active comets C/1983 H1 IRAS-Araki- Alcock, 26P/Grigg-Skjellerup, and C/1996 B2 Hyakutake [29], which have lower values of the polarization ratio (μ < 0.5) and low albedos (n_{OC} < 0.1).

Event though, radar characteristics of inactive comet nuclei are still not studied, but recent studies have shown that some asteroids with noncomet like orbits may have a comet like dust tail [30]; however, up to now, there is no evidence that asteroid 1998 WT24 is a nucleus of an inactive comet.

Having measurements of the radar albedo and the surface roughness we can make several conclusions about the properties of the surface material; however, these conclusions have limited reliability owing to the absence of a satisfactory theory of diffuse scattering of radio waves. If we assume that the asteroid is a solid rock, the radar albedo may be related to the Fresnel coefficient R via the formula: $n_{OC} = gR$, where coefficient g depends on the asteroid orientation, shape, and roughness of the relief.

Then, assuming that, for a sphere with the scattering law $\cos^n\theta$, coefficient g can be found from the relationship [31]:

$$g = (n+2)/(n+1)$$

We find that, for the determined value of the roughness, n = 1.8, coefficient g is approximately 1.36 and R ≈ 0.11. It allows us to estimate ground density as ~2.3 g/cm^3 using equation from [32] and the asteroid mass ~$7.8 \cdot 10^{10}$ kg. Taking into account assumption that asteroid surface porosity are the same as a porosity of the upper layer of lunar regolith and using estimated ground density we can estimate the particle density as ~4.5 g/cm^3. Assuming dry rocks as a surface of asteroid 1998 WT24, the ground permittivity is ~4.5; however, this value is an averaged parameter for a surface layer whose depth is equal to the radio-wave penetration depth. In addition, it should be taken into account that, in the centimeter wave band, materials with different ground permittivity may have close scattering patterns. In case of the asteroid surface is formed with ice, Fresnel reflection coefficient R can be found from the relationship [33]:

$$R = \left(\frac{1-\sqrt{\varepsilon}}{1+\sqrt{\varepsilon}}\right)^2,$$

where permittivity ε of the surface is connected with the permittivity of ice via the Rayleigh relationship. Assuming that the porosity of ice is ~50%, we obtain the following values of the permittivity ~1.8 and the reflection coefficient ~0.02. In this case, it is difficult to estimate the asteroid mass, because radio waves are reflected from a thin surface ice layer and it is impossible to find the density of the asteroid's non-ice components, although the contribution from these components to the asteroid mass is dominant.

Scattering Properties of Asteroid

Our technique of scattering properties construction described in [17] allows not only to get an equatorial hull of the asteroid but also to estimate some of the scattering properties of its surface using another representation of the spectra intensity as a convex outline with scattering information overlaid on it. A scattering image constructed by our method has the following properties: if an object represents an aggregation of local reflection points with distances between points being substantially larger than their linear dimensions, then a scattering image will have a patchy structure that corresponds to distribution of the points[35]; if distances between local scattering points are small, as in the case of a rough ellipsoid [17], brightness distribution of a scattering image is determined by the scattering law of the object surface. In case of rough asteroid surface with increased or decreased reflectivity regions, a scattering image will have regions with high or low brightness, respectively.We choose all 65 spectra from sessions 2, 3, 4 and only 36 spectra from sessions 1, 6 with the data needed for the whole hull reconstruction. Spectra from session 5 are not used because of very low reliability. Reconstruction time between the spectra was about ~1.5 min ; it corresponds to the rotation of the asteroid ~0.04 rad. A scattering image of asteroid 1998 WT24 shows considerable structures, identical in both OC and SC polarizations (Figure 3).

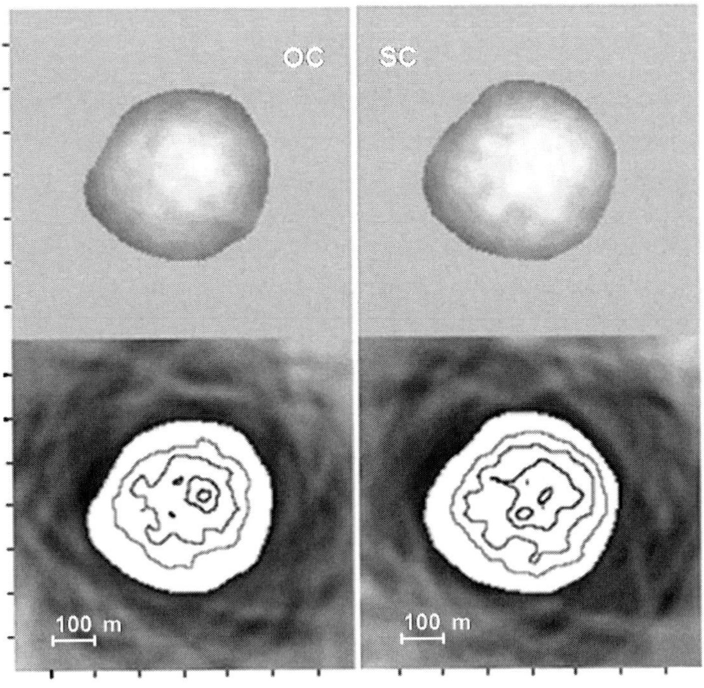

Figure 3. Scattering images of asteroid 1998 WT24 obtained for OC (left) and SC (right) polarizations.

On the top images, a dark-gray color we interpret as a concavity while group of radar-bright pixels we interpret as a raised feature with the evidence of high scattering. The bottom images provide more convenient information representing the same intensity of pixels in the form of contour lines corresponding to identical brightness of the top image.

The contour lines within hull correspond to four values of brightness: 0.97, 0.9, 0.8 and 0.6 of the maximum value of brightness. The brightness outside hull are characterized the accidental nature of a noise.

Certainly, our images demonstrate only integral scattering features and have less surface details as the radar image obtained by Ostro[5]. However, comparison of scattering features against the delay–Doppler images of WT24 presented in [5] shows that our images does not only provide information about asteroid hull, but also indicates areas with features on the surface like peaks or edges whose scattering features are unique while small details of the relief are missed.

CONCLUSION

Analysis of radar data of 1998 WT24 shows that variations in the integral energy of the echo-signal power spectra as functions of the asteroid aspect angle are small. The energy difference is characterized by a factor of 1.2 for the SC polarization and by a factor of 1.5 for the OC polarization. At certain aspect angles, the power spectra contain two or three local energy maxima indicating the presence of large-scale forms of the relief. The polarization ratio of the asteroid is ~1.05.

The maximum, minimum, and average asteroid diameters D_{max} = 430±10 m, D_{min} = 380±10 m, and D_{avg} = 405±10 m almost the same on both polarizations. The asteroid's radar cross section is ~0.021 km^2, the radar albedo is ~0.15, and the optical albedo is ~0.41 (assuming H=18.54). A low albedo and a near-unit polarization ratio indicate that the asteroid has a strongly rough surface.

The result of simulation has shown that the degree of the asteroid roughness is ~1.8; however, the surface is nonuniformly rough. Even though, the characteristics of this asteroid are similar to the characteristics of the Jovian ice satellites, it is impossible to unambiguously conclude that the asteroid consists of ice. If we assume that asteroid 1998 WT24 is a solid rock, the density of the surface ground may be ~2.3 g/cm^3, the ground permittivity may be ~4.5, and the asteroid mass may be ~$7.8 \cdot 10^{10}$ kg. Images of scattering properties show hull of the asteroid as well as indicate any scattering features on the surface. The result is in agreement with other radar and optical experiments.

REFERENCES

[1] Harris, A.W.; Mueller, M., Delbó, M., Bus, S.J. (2007). Physical characterization of the potentially hazardous high-albedo Asteroid (33342) 1998 WT24 from thermal-infrared observations. *Icarus* 188, 414–424.

[2] Kiselev, N.N., Rosenbush, V.K., Jockers, K., Velichko, F.P., Shakhovskoj, N.M., Efimov, Y.S., Lupishko, D.F., Rumyantsev, V.V., (2002). Polarimetry of near-Earth Asteroid 33342 (1998 WT24). In: Warmbein, B. (Ed.), *Proc. of the Conference: Asteroids, Comets, Meteors,* ACM 2002. ESA SP-500. ESA, Noordwijk, The Netherlands, pp. 887–890.

[3] Krugly, Y.N., Belskaya, I.N., Chiorny, V.G., Shevchenko, V.G., Gaftonyuk, N.M., (2002). CCD photometry of near-Earth asteroids in 2001. In: Warmbein, B. (Ed.), *Proc. of the Conference: Asteroids, Comets, Meteors,* ACM 2002. ESA SP-500. ESA, Noordwijk, The Netherlands, pp. 903–906.

[4] Lazzarin, M., Marchi, S., Barucci, M.A., Di Martino, M., Barbieri, C., (2004). Visible and near-infrared spectroscopic investigation of near-Earth objects at ESO: First results. *Icarus* 169, 373–384.

[5] Busch, M. W., Benner, Lance A. M., Ostro, S. J., Giorgini, J. D., Jurgens, R. F., Rose, R., Scheeres, D. J., Magri, C, Margot, J., Nolan, M.C. , Hine, A. A. (2008) Physical properties of near-Earth Asteroid (33342) 1998 WT24. *Icarus,* 195 (2). pp. 614-621.

[6] M. Di Martino, S. Montebugnoli, G. Cevolani, S. Ostro, A. Zaitsev, S. Righini, L. Saba, S. Poppi, M. Delbт, A. Orlati, G. Maccaferri, C. Bortolotti, A. Gavrik and Y. Gavrik. (2004) Results of the first Italian planetary radar experiment. *Planetary and Space Science,* Volume 52, Issue 4, p. 325.

[7] A. L. Zaitsev, M di Martino, A A. Konovalenko, S Montebugnoli, S P. Ignatov, Y F. Kolyuka, A S. Nabatov, I S. Falkovich, A L. Gavrik, Y A. Gavrik, C Bortolotti, A Cattani, A Maccaferri, G Maccaferri, M Roma, M Delbo, L Saba, T Afanas'eva, T Gridchina, S Righini. Radar detection of NEA 33342 (1998 WT24) with EVPATORIA => MEDICINA system

at 6 cm. *Proceedings of the ACM 2002,* Berlin, ESA-SP-500, pp. 883-886.

[8] Yu. A. Gavrik and A. L. Gavrik Radar Characteristics of Asteroid 33342 (1998 WT24) *Journal of Communications Technology and Electronics,* 2008, Vol. 53, No. 10, pp. 1177–1185.

[9] S. L. Marple, Digital Spectral Analysis with Application (Prentice-Hall, Englewood Cliffs, N.J., 1987; Mir, Moscow, 1990).

[10] S. J. Ostro. *Rev. Mod. Phys.* 65, 1235 (1993).

[11] S. J. Ostro. Hazards Due to Comets and Asteroids, Ed. by T. Gehrels and M. S. Matthews (Univ. Arizona, Tucson, 1994), p. 259.

[12] B. Zellner, Asteroids, Ed. by T. Gehrels (Univ. Arizona, Tucson, 1979), p. 783.

[13] M. Delbo, *Doctoral Dissertation* (Freie Universitat Berlin, Berlin, 2004).

[14] Data Base of Physical and Dynamical Properties of Near Earth Asteroids. http://earn.dlr.de/nea/033342.htm.

[15] S. J. Ostro. Planetary Radar. *Encyclopedia of the Solar System* (Academic, New York, 1999).

[16] S. J. Ostro, R. D. West, M. A. Janssen et al., *Icarus* 183, 479 (2006).

[17] A. L. Gavrik, Yu. A. Gavrik, V. N. Sekistov, and V. A. Sorokin. The Use of Radio Images for Analyzing the Results of Narrow-Band Radio Location of Asteroids. *Cosmic Research,* 2009, Vol. 47, No. 3, pp. 208–216.

[18] R. F. Jurgens, *Icarus* 49, 97 (1982).

[19] S. J. Ostro, R. S. Hudson, K. D. Rosema et al., *Icarus* 137, 122 (1999).

[20] Bus, S.J., Villas, F., Barucci, M.A., 2002. Visible-wavelength spectroscopy of asteroids. In: Asteroids III. Univ. of Arizona Press, Tucson, AZ, pp. 169–182.

[21] NEA Circular Polarization Ration Ranking. http://echo.jpl.nasa.gov /~lance/asteroid-radar-properties/nea.sc-oc.html.

[22] S. J. Ostro, D. B. Campbell, R. A. Simpson et al., *J. Geophys. Res.* 97, 18277 (1992).

[23] D. O. Muhleman, B. J. Butler, A. W. Grossman, and M. A. Slade, *Science* (Washington, D.C.) 253, 1508 (1991).

[24] M. A. Slade, B. J. Butler, and D. O. Muhleman, *Science* 258, 635 (1992).

[25] B. A. Campbell, R. E. Arvidson, and M. K. Shepard, *J. Geophys. Res.* 98, 17099 (1993).

[26] K. A. Tryka and D. O. Muhleman, *J. Gcophys. Res.* 97, 379 (1992).

[27] J. K. Harmon and S. J. Ostro, *Icarus* 62, 110 (1985).
[28] N. J. S. Stacy, D. B. Campbell, and P. G. Ford, *Science* 276, 1527 (1997).
[29] L. A. M. Benner, S. J. Ostro, J. D. Giorgini, et al., *Icarus* 130, 296 (1997).
[30] H. H. Hsieh, D. C. Jewitt, and Y. R. Fernandez, *Astron. J.* 127, 2997 (2004).
[31] L. A. M. Benner, R. S. Hudson, S. J. Ostro, et al., *Icarus* 139, 309 (1999).
[32] G. H. Heiken, D. T. Vaniman, and B. M. French, *Lunar Sourcebook* (Cambridge Univ. Press, New York, 1991).
[33] G. J. Black, D. B. Campbell, and S. *J. Ostro, Icarus* 151, 160 (2001).
[34] Zaitsev, A. L., S. J. Ostro, S. P. Ignatov, D. K. Yeomans, A. G. Petrenko, D. Choate, O. K. Margorin, R. A. Cormier, V. V. Mardyshkin, R. Winkler, O. N. Rghiga, R. F. Jurgens,V. A. Shubin, J. D. Giorgini, A. P. Krivtsov, K. D. Rosema, Y. F. Koluka, M. A. Slade, A. L. Gavrik, V. B. Andreev, D. V. Ivanov, P. S. Peshin, Y. Koyama, M. Yoshikava, A. *Nakamura. Planetary Space Sci.* 45, 771-778. (1997).
[35] V.N. Sekistov, A.L. Gavrik, V.J Andreev, S.G Khaplanov, V.A. Sorokin, A.A. Kotov, J.A. Gavrik. Reconstruction of radio images of near Earth satellites from narrow band radar data. *Journal of Communications Technology and Electronics.* 2000. V. 45, No 7, pp. 830-836.

In: Meteorites and Asteroids
Editors: A. Dementieva et al.
ISBN 978-1-61942-115-8

Chapter 5

ASTEROIDS DIMENSIONS AND THE TRUNCATED PARETO DISTRIBUTION

Lorenzo Zaninetti **[*]** ***and Mario Ferraro*** **[†]**
Dipartimento di Fisica Generale
Via Pietro Giuria 1, 10125
Torino, Italy

Abstract

In this chapter first the statistics of the standard and truncated Pareto distributions are derived and used to fit empirical values of asteroids diameters from different families, namely, Koronis, Eos and Themis, and from the Astorb database. A theoretical analysis is then carried out and two possible physical mechanisms are suggested that account for Pareto tails in distributions of asteroids diameter.

PACS 96.30.Ys.

Keywords: Asteroids.

1. Introduction

The interest in the study of asteroids in the inner solar system lays in their connection with the formation of planets and their temporal evolution.

Among others, studies on asteroids formations and evolutions involve

[*]Email address: zaninetti@ph.unito.it
[†]Email address: ferraro@ph.unito.it

1. the effects of asteroid collisional history on sizes and spins of present-day objects [1],

2. realistic collisional scaling laws and the implications of including observables, such as collisional produced families, in constraining the collisional history of main-belt asteroids [2],

3. the creation of a model of the main asteroid belt whose purpose is to describe distribution of size size frequency of asteroids and simulate their number [3].

4. temporal evolution for 4.2 Myr of test particles, which were initially placed on a perfectly rectangular grid and subjected to gravitational interactions with the Sun and five planets, from Mars to Neptune, see [4].

On the other hand, it has been shown that experimental observations can be fitted with a differential size distribution

$$n = dN/dD = n(D) \propto D^{-\alpha}, \tag{1}$$

where D is the diameter in Km, α the exponent of the inverse power law and n the number of asteroids comprised between D and $D + dD$. Measurements of the properties of 13,000 asteroids detected by the Sloan Digital Sky Survey (SDSS) present a differential size distribution that for $D \geq 5Km$ is $n \propto D^{-4}$ and for $D \leq 5Km$ is $n \propto D^{-2.3}$, see [4].

The ongoing simulations as well the observations require a careful analysis of the Pareto distribution [5, 6] and the truncated Pareto distribution [7, 8, 9]. This paper presents in Section 2. a comparison between the Pareto and the truncated Pareto distributions. In Section 3. the theoretical results are applied to the distribution of the radius of asteroids. Two physical mechanisms that produces a Pareto type distribution for diameters are presented in Section 4.

2. Statistical Distribution

Let X be a random variable taking values x in the interval $[a, \infty]$, $a > 0$. The probability density function (in the following pdf) named Pareto is defined by [6]

$$f(x; a, c) = ca^c x^{-(c+1)}, \tag{2}$$

$c > 0$, and the Pareto distribution functions is $F(x : a, c) = 1 - a^c x^{-c}$.

An upper truncated Pareto random variable is defined in the interval $[a, b]$ and the corresponding pdf is

$$f_T(x; a, b, c) = \frac{c a^c x^{-(c+1)}}{1 - \left(\frac{a}{b}\right)^c}, \tag{3}$$

[9] and the truncated Pareto distribution function is

$$F_T(x; a, b, c) = \frac{1 - \left(\frac{a}{x}\right)^c}{1 - \left(\frac{a}{b}\right)^c}. \tag{4}$$

Momenta of the truncated distributions exist for all $c > 0$. For instance, the mean of $f_T(x; a, b, c)$ is, for $c \neq 1$ and $c = 1$, respectively,

$$\langle x \rangle = \frac{ca}{c-1} \frac{1 - \left(\frac{a}{b}\right)^{c-1}}{1 - \left(\frac{a}{b}\right)^c}, \quad \langle x \rangle = \frac{c a^c}{1 - \left(\frac{a}{b}\right)^c} \ln \frac{b}{a}. \tag{5}$$

Similarly, if $c \neq 2$, the variance is given by

$$\sigma^2 = \frac{ca^2}{(c-2)} \frac{1 - \left(\frac{a}{b}\right)^{c-2}}{1 - \left(\frac{a}{b}\right)^c} - \langle x \rangle^2, \tag{6}$$

whereas for $c = 2$

$$\frac{c a^c}{1 - \left(\frac{a}{b}\right)^c} \ln \frac{b}{a} - \langle x \rangle^2. \tag{7}$$

In general the $n - th$ central moment is

$$\int_a^b (x - \langle x \rangle)^n f_T(x) dx =$$

$$= (-\langle x \rangle)^n a^{-c} {}_2F_1 \left(-c, -n; 1-c; \frac{a}{\langle x \rangle} \right) \left((a^c)^{-1} - (b^c)^{-1} \right)^{-1}$$
$$- (-\langle x \rangle)^n b^{-c} {}_2F_1 \left(-c, -n; 1-c; \frac{b}{\langle x \rangle} \right) \left((a^c)^{-1} - (b^c)^{-1} \right)^{-1}, \tag{8}$$

where ${}_2F_1(a, b; c; z)$ is a regularized hypergeometric function, see [10, 11, 12]. An analogous formula based on some of the properties of the incomplete beta function, see [13] and [14], can be found in [15]. The median m of the Pareto distribution is

$$m = 2^{c^{-1}} a, \tag{9}$$

and the median of truncated Pareto m_T

$$m_T = a2^{c^{-1}} \left(a^c b^{-c} + 1\right)^{-c^{-1}} . \tag{10}$$

Parameters of the truncated Pareto pdf from empirical data can be obtained via the maximum likelihood method; explicit formulas for maximum likelihood estimators (MLE) are given in [7], and for the more general case in [9], whose results we report here for completeness.

Consider a random sample $\mathcal{X} = x_1, x_2, \ldots, x_n$ and let $x_{(1)} \geq x_{(2)} \geq \cdots \geq x_{(n)}$ denote their order statistics so that $x_{(1)} = \max(x_1, x_2, \ldots, x_n)$, $x_{(n)} = \min(x_1, x_2, \ldots, x_n)$.

The MLE of the parameters a and b are

$$\tilde{a} = x_{(n)}, \quad \tilde{b} = x_{(1)}, \tag{11}$$

respectively, and $\tilde{c}$ is the solution of the equation

$$\frac{n}{\tilde{c}} + \frac{n(\frac{x_{(n)}}{x_{(1)}})^{\tilde{c}} \ln\left(\frac{x_{(n)}}{x_{(1)}}\right)}{1 - \left(\frac{x_{(n)}}{x_{(1)}}\right)^{\tilde{c}}} - \sum_{i=1}^{n} \left[\ln x_i - \ln x_{(n)}\right] = 0, \tag{12}$$

[9].

There exists a simple test to see whether a Pareto model is appropriate [9]: the null hypothesis $H_0 : \nu = \infty$ is rejected if and only if $x_{(1)} < [nC/(-\ln q)]^{1/c}$, $0 < q < 1$, where $C = a^c$. The approximate p-value of this test is given by $p = \exp\left\{-nCx_{(1)}^{-c}\right\}$, and a small value of p indicates that the Pareto model is not a good fit.

Given a set of data is often difficult to decide if they agree more closely with f or f_T, since, in the interval $[a, b]$, they differ only or a multiplicative factor $1 - (a/b)^c$, that approaches 1 even for relatively small values of c if the interval $[a, b]$ is not too small. For this reason, rather than f and f_T, the distributions $P(X > x)$ and $P_T(X > x)$ are used, often called survival functions, that are given respectively by

$$P(X > x) = S(x) = 1 - F(x; a, c) = a^c x^{-c} \tag{13}$$

and

$$P_T(X > x) = S_T(x) = 1 - F_T(x; a, b, c) = \frac{ca^c\left(x^{-c} - b^{-c}\right)}{1 - \left(\frac{a}{b}\right)^c} . \tag{14}$$

The Pareto variate X can be generated by

$$X : a, c \sim a\,(1 - R)^{-\frac{1}{c}}\ , \tag{15}$$

and the truncated Pareto variate X_T by

$$X_T : a, b, c \sim a\left(1 - R\left(1 - \left(\frac{a}{b}\right)^c\right)\right)^{-\frac{1}{c}}\ , \tag{16}$$

where R is the unit rectangular variate.

3. Application to the Asteroids

We have tested the hypothesis that diameters of asteroids follows a Pareto distribution by by considering different families of asteroids, namely, Koronis, Eos and Themis.

The sample parameter of the families are reported in Table 1, Table 2 and Table 3, whereas Figure 1, Figure 2, Figure 3 report the graphical display of data and the fitting distributions.

Table 1. *Coefficients of diameter distribution of the Koronis family. The parameter c is derived through MLE and $p = 0.033$.*

a [km]	b [km]	c	n	$P(X > x)$
25.1	44.3	3.77	29	truncated Pareto
25.1	∞	5.04	29	Pareto

Table 2. *Coefficients of diameter distribution of the Eos family. The parameter c is derived through MLE and $p = 0.681$.*

a [km]	b [km]	c	n	$P(X > x)$
30.1	110	3.80	53	truncated Pareto
30.1	∞	3.94	53	Pareto

In case of the Koronis family P_T fits the data better than P and indeed $p = 0.039$ is correspondingly small, whereas for the Eos family, P performs slightly better than P_T ($p = 0.68$), and the estimated of c are very closed in

Table 3. *Coefficients of diameter distribution of the Themis family. The parameter c is derived through MLE and* $p = 0.67$.

$a\ [km]$	$b\ [km]$	c	n	$P(X > x)$
35.3	249	2.5	53	truncated Pareto
35.3	∞	2.6	53	Pareto

both cases. Finally in the third case, the Themis family, the two distributions are the same, due to the fact that the ratio $a/b = 0.14$ is small.

Another interesting catalog is the Asteroid Orbital Elements Database (Astorb) which is visible at http://vizier.u-strasbg.fr/; the sample parameter of the asteroids with diameter $> 90\ Km$ is reported in Table 4 and the fitting survival function in Figure 4.

Table 4. *Coefficients of diameter distribution of the asteroids extracted from Astorb database with diameter* $> 90\ Km$. *The parameter c is derived through MLE and* $p = 0.53$.

$a\ [km]$	$b\ [km]$	c	n	$P(X > x)$
90.59	848.4	2.71	272	truncated Pareto
90.59	∞	2.75	272	Pareto

4. Simulating Pareto Tails

Two models that explains the Pareto tails are presented. The first analyzes the possibility that the asteroids are formed from smaller bodies and the second one introduces a fragmentation model at the light of the Voronoi diagrams.

4.1. Accretion

As a simple example of how a distribution with power can be generated, consider the growth of a primeval nebula via accretion, that is the process by which nebulae "capture" mass. We start by considering an uniform pdf for the initial mass of N primeval nebulae, m, in a range $m_{min} < m \leq m_{max}$. At each interaction the i-th nebula has a probability λ_i to increase its mass m_i

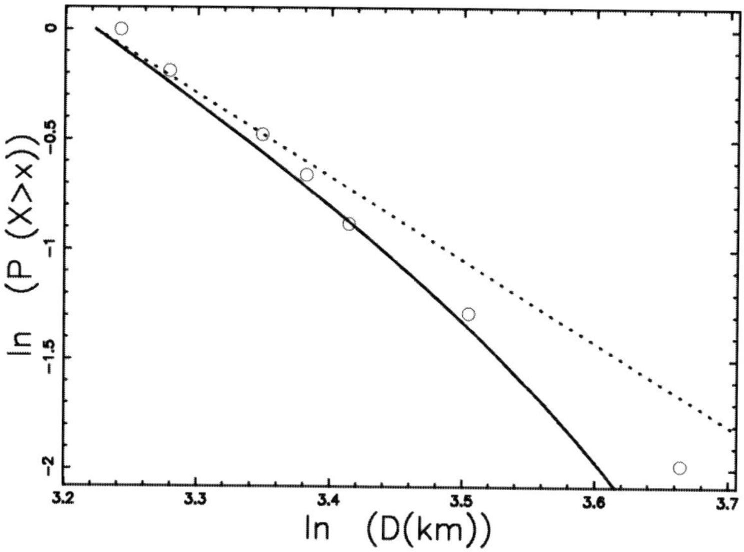

Figure 1. ln–ln plot of the survival function of diameter distribution of the Koronis Family: data (empty circles), survival function of the truncated Pareto pdf (full line) and survival function of the Pareto pdf (dotted line). A complete sample is considered with parameters as in Table 1.

that is given by

$$\lambda_i = (1 - \exp(-akm_i)), \tag{17}$$

where ak is a parameter of the simulation; thus more "massive" nebulae are more likely to grow, via accretion. The quantity of which the primeval nebula can grow varies with time, to take into account that the total mass available is limited,

$$\delta m(t) = \delta m(0) \exp(-t/\tau), \tag{18}$$

where $\delta m(0)$ represents the maximum mass of exchange and τ the scaling time of the phenomena. The simulation proceeds as follows: a number r, is randomly chosen in the interval $[0, 1]$ for each nebula, and, if $r < \lambda_i$, the mass m_i is increased by $\delta m(t)$, where t denotes the iteration of the process. The process proceed in parallel: at each temporal iteration all the primeval nebulae are considered.

Results of the simulations have been fitted with both Pareto survival distributions (see Figure 5).

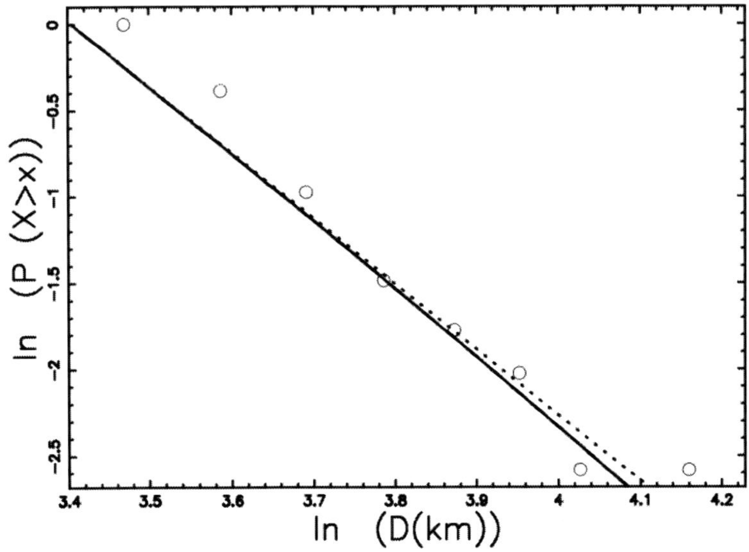

Figure 2. ln–ln plot of the survival function of diameter distribution of the Eos Family: data (empty circles), survival function of the truncated Pareto pdf (full line) and survival function of the Pareto pdf (dotted line). A complete sample is considered with parameters as in Table 2.

4.2. Fragmentation

The distribution of fragments size arising from breaking of material is still subject of research as it depends on the actual fragmentation process. The first model was developed in [16]: it assumes a time dependent probability of fracture of a ring after a critical strain has been reached in the material. The resulting distribution show that the frequency of occurrence of fragments masses follows a cumulative distribution of the form:

$$\frac{N_m}{N} = e^{-\sqrt{\frac{m}{\mu}}} , \tag{19}$$

where N_m is the number of fragments each of whose mass is greater or equal to m, N is the total number of fragments,

$$\mu = \frac{\overline{m}}{2} , \tag{20}$$

and $\overline{m}$ is the averaged mass of a fragment.

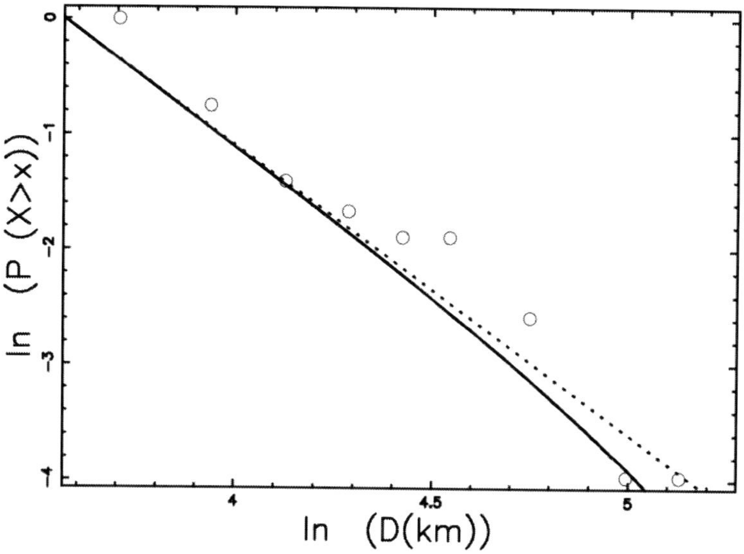

Figure 3. ln–ln plot of the survival function of diameter distribution of the Themis Family: data (empty circles), survival function of the truncated Pareto pdf (full line) and survival function of the Pareto pdf (dotted line). A complete sample is considered with parameters as in Table 3.

4.2.1. Fractal distribution of fragments size

In order to generate cells resulting in a fractal distribution of their volumes the following method can be adopted, which generalizes the procedure presented in [17].

Consider a domain $\mathcal{D}$ subdivided into N_0 cubic cells with a linear dimension l that, that in the following, will be referred as zero-order cells. A zero-order cell is divided into k^3 smaller cubes called zero-order elements, with linear dimension l/k and volumes

$$V_1 = \frac{V_0}{k^3}, \tag{21}$$

where V_0 is the volume of zero-order cell. If P is the probability of a zero-order cell to be fragmented, the number N_1 of zero-order elements generated by fragmentation is given by

$$N_1 = P \cdot k^3 N_0, \tag{22}$$

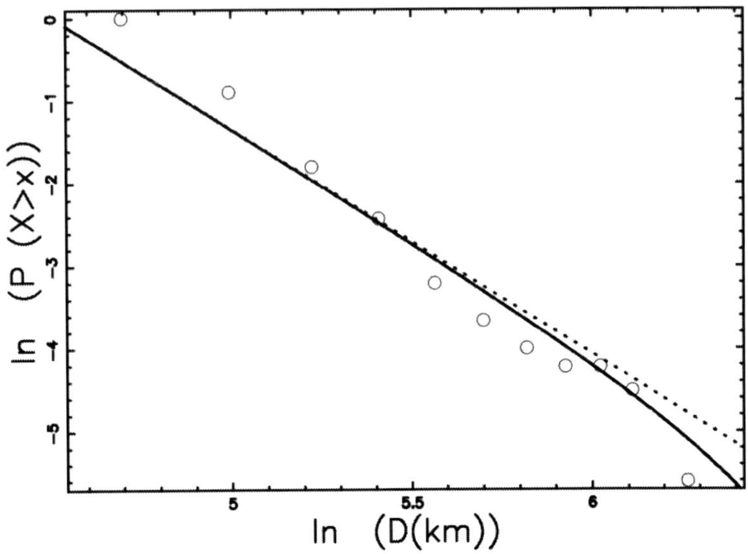

Figure 4. ln–ln plot of the survival function of diameter distribution of the asteroids extracted from Astorb database with diameter $> 90\ Km$, data (empty circles), survival function of the truncated Pareto pdf (full line) and survival function of the Pareto pdf (dotted line). A complete sample is considered with parameters as in Table 4.

and the number N_{0a} of zero-order cells that have been not fragmented is

$$N_{0a} = (1 - P)N_0. \tag{23}$$

Each zero-order element now becomes a first-order cell that can be fragmented into first-order elements of volumes

$$V_2 = \frac{V_0}{(k^3)^2} \tag{24}$$

and the number of fragmented first-order elements is

$$N_2 = Pk^3 N_1 = (k^3 P)^2 N_0. \tag{25}$$

The number N_{1a} of first order cells that have not been fragmented is given by

$$N_{1a} = k^3 P(1 - P)N_0. \tag{26}$$

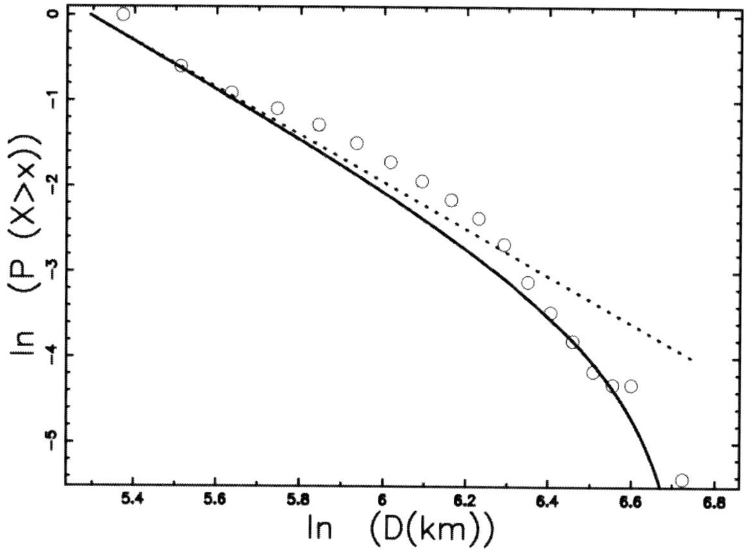

Figure 5. log–log plot of the survival function of the diameter distribution for the primeval nebula. The truncated Pareto parameters are $c = 2.75$ and $p = 0.00026$.

Now first-order elements can be considered second-order cells and the procedure repeats itself. The volume of the nth-order cell V_n is

$$V_n = \frac{V_0}{k^{3n}}, \tag{27}$$

and, after fragmentation, the number of nth-order cells N_{na} is

$$N_n = (Pk^3)^n N_{0a} = (Pk^3)^n (1-P) N_0. \tag{28}$$

Taking the natural logarithm Eqs. (27) and (28) leads to

$$\ln \frac{V_n}{V_0} = -n \ln(k^3), \tag{29}$$

$$\ln \frac{N_{na}}{N_{0a}} = -n \ln(Pk)^3 . \tag{30}$$

From Eqs. (30) and (30) it is straightforward to obtain, by elimination of n:

$$\frac{N_{na}}{N_{0a}} = \left[\frac{V_n}{V_0}\right]^{-\frac{\ln[Pk^3]}{\ln[k^3]}}, \tag{31}$$

that is a fractal distribution with dimension D given by

$$D = \frac{3\ln(Pk^3)}{\ln(k^3)} . \tag{32}$$

Now we can consider the center of cells, of any order, as seeds for the generation of a Voronoi diagram as shown in see Figure 6.

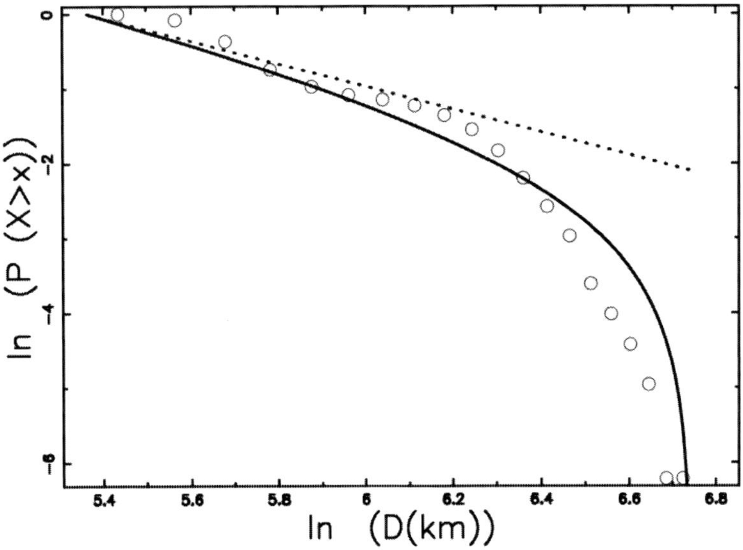

Figure 6. log–log plot of the survival function of the diameter distribution as given by the Voronoi Diagrams in presence of 1000 fractal seeds. The parameters of the simulation are $k = 2$, $P = 0.92$ and therefore $D_{fr} = 2.80$. The truncated Pareto parameters are $c = 1.52$ and $p = 0$.

5. Conclusions

In this chapter statistical parameters for the truncated Pareto distribution, namely average, variance, median and $n - th$ central moment have been calculated. Furthermore, also distribution function and survival function have been derived. These quantities allow to fit the various families of asteroids and the Astorb database which are characterized by a finite rather than infinite maximum diameter. Two possible simulations are suggested to produce Pareto tails. The first one results from a simple growth process, in which the increase of the state variable (here mass) depends on the values taken in the previous state. The second one is a fragmentation process given by 3D Voronoi volumes with a fractal distribution of seeds.

References

[1] D. R. Davis, S. J. Weidenschilling, P. Farinella, P. Paolicchi, R. P. Binzel, Asteroid collisional history - Effects on sizes and spins, in: R. P. Binzel, T. Gehrels, & M. S. Matthews (Ed.), Asteroids II, 1989, pp. 805–826.

[2] D. R. Davis, D. D. Durda, F. Marzari, A. Campo Bagatin, R. Gil-Hutton, Asteroids III (2002) 545–558.

[3] E. F. Tedesco, A. Cellino, V. Zappalá, AJ 129 (2005) 2869–2886.

[4] T. A. Michtchenko, D. Lazzaro, J. M. Carvano, S. Ferraz-Mello, MNRAS 401 (2010) 2499–2516.

[5] V. Pareto, Cours d' economie politique, Rouge, Lausanne, 1896.

[6] M. Evans, N. Hastings, B. Peacock, Statistical Distributions - third edition, John Wiley & Sons Inc, New York, 2000.

[7] A. Cohen, B. Whitten, Parameter Estimation in reliability and Life Span Models, Marcel Dekker, New York, 1988.

[8] D. Devoto, S. Martinez, Mathematical Geology 30 (6) (1998) 661 – 673.

[9] I. Aban, M. Meerschaert, A. Panorska, Journal of the American Statistical Association 101 (2006) 270–277.

[10] M. Abramowitz, I. A. Stegun, Handbook of mathematical functions with formulas, graphs, and mathematical tables, Dover, New York, 1965.

[11] D. von Seggern, CRC Standard Curves and Surfaces, CRC, New York, 1992.

[12] W. J. Thompson, Atlas for computing mathematical functions, Wiley-Interscience, New York, 1997.

[13] I. Gradshteyn, I. Ryzhik, Table of Integrals, Series, and Products, Academic Press, San Diego, 2000.

[14] A. Prudnikov, O. Brychkov, Y. Marichev, Integrals and Series, Gordon and Breach Science Publishers, Amsterdam, 1986.

[15] M. Masoom Ali, S. Nadarajah, Computer Communications 30 (2006) 1–4.

[16] N. F. Mott, Royal Society of London Proceedings Series A 189 (1947) 300–308.

[17] Turcotte, D. L., Fractals and chaos in geology and geophysics., Cambridge University Press., Cambridge, 1993.

In: Meteorites and Asteroids
Editors: A. Dementieva et al.
ISBN: 978-1-61942-115-8

Chapter 6

HILDA ASTEROIDS IN THE JUPITER NEIBORHOOD

Romina P. Di Sisto and Adrián Brunini
Univ. Nacional de La Plata, Argentina

Abstract

Hilda asteroids are objects that orbit the Sun at a distance where the orbital period is exactly 2/3 of the orbital period of Jupiter. Previous analytical and numerical studies have focused their attention to the great dynamical stability within the orbital region where the Hilda asteroids are found. However the group is indeed dispersing. In this paper we are going to review the contribution of Hilda asteroids to the Jupiter neiborhood. Based on simulations of long term dynamical evolution of escaped Hilda asteroids, it is find that 8% of the particles leaving the resonance end up impacting Jupiter. Also they are the main source of small craters on the Jovian satellite system, overcoming the production rate by comets and Trojan asteroids. Almost all the escaped Hildas pass through the dynamical region occupied by JFCs, and the mean dynamical life time there is 1.4×10^6 years. $\sim 14\%$ of JFCs with $q > 2.5$ A.U. could be escaped Hilda asteroids. We analyzed also the possibility that the Shoemaker - Levy 9 was an Hilda asteroid.

1. Introduction

Asteroids are small solid bodies that orbit the Sun. In general they are rocky, some ones have a metal composition, and others are rich in water ice, other volatiles and organic materials. To this last group belong the Hilda asteroids.

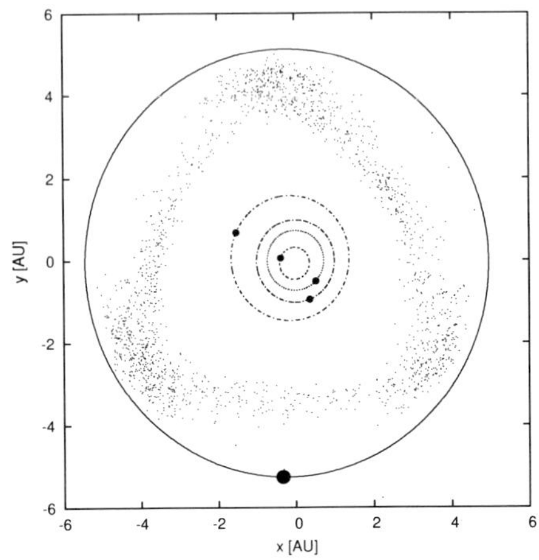

Figure 1. Instantaneous positions of Hilda asteroids (dots) in the ecliptic coordinate system, projected onto the plane of the ecliptic for the epoch UT 0 h 27 October 2007 . The orbits of Jupiter and the terrestrial planets are drawn for scale. The black points represent the location of the planets from Mercury to Jupiter, for the Hildas osculating epoch.

They are placed at $\sim$ 4 AU in the $3:2$ mean motion resonance with Jupiter, namely in the time that Jupiter completes two periods an Hilda asteroid completes three. It is a numerous group of asteroids that make evident the protecting mechanism present in this resonance. The angle $\varphi = 3\lambda_J - 2\lambda - w$, where λ_J and λ are the mean longitudes of Jupiter and the asteroid respectively and w is the longitude of perihelion of the asteroid, is called critical angle and librates around zero. Then, when the asteroid is in conjunction with Jupiter ($\lambda = \lambda_J$), it is near perihelion ($\lambda \sim w$) and far from the planet. The configuration of Hildas with respect to Jupiter, and the mentioned protecting mechanism can be seen in Fig. 1. There it is plotted the instantaneous positions of Hilda asteroids obtained from the ASTORB database of the osculating elements for asteroids (maintained by E. G. Bowell of Lowell Observatory), projected onto the plane of the ecliptic. There it is also shown the location of the planets from Mercury to Jupiter for the Hildas osculating epoch, and their orbits.

The central zone of the 3:2 resonance has a great dynamical stability, where an asteroid can last for the age of the solar system (Nervorný et al. 1997, Ferraz Mello et al. 1998). However, the dynamical evolution of Hildas and mainly their colisional activity can remove them from this protected zone and they can reach quickly the unstable regions that surround the stable zone. Then they can reach a very chaotic boundary, where the characteristic permanence times are very short and they will be quickly ejected.

In this paper we are going to review the contribution of Hilda asteroids to the Jupiter neiborhood. We will see the dynamical evolution of Hilda asteroids that are recently removed from their stable orbits, their relation to Jupiter Family Comets (JFC) and the contribution to the cratering rate on the major satellites of Jupiter.

The general dynamics of Hildas recently removed from the resonance is addressed in section 3.

Comets, in particular Jupiter Family Comets (JFCs), are minor bodies whose dynamical evolution is mainly controlled by Jupiter and so they pass through the Jupiter's neiborhood. JFCs come from the Trans Neptunian region (Fernández, 1980), they have unstable orbits and a short lifetime. They are formed by a solid nucleus consisting of intimately mixed ices and dust. When they approach to the Sun, the ices sublimate forming the coma and the tail. Hilda asteroids are in the external zone of the asteroid belt and they could also contain volatile ices, like comets. In fact they have D and P taxonomic class that can be considered as spectral analogous of comets. However Hilda asteroids couldn't approach enough to the Sun to have the possibility of sublimate their volatiles. They could be dormant comets, this is a comet nucleus with no detectable activity, but if delivered to lower semi major axes, it could be reactivated, according to the definition by Hartmann et al. (1987). After escape from the resonance, Hildas follow dynamical routes that are controlled by Jupiter (Di Sisto et al. 2005), they can reach low perihelion distances and then they could have the possibility of sublimate their volatiles. JFCs are mainly found in the same zone of escaped Hildas (Di Sisto, et al. 2005). Also there are JFCs associated with the Hilda asteroid region (Toth, 2006) and with its boundaries (Kresák, 1979). This last group, called the quasi-Hilda group of comets, is very important because it could give the keys of the relation between asteroids and comets.

The relation between asteroids and comets is a vastly studied subject but not an already closed one. There are many questions that continue still open and in discussion. In particular the relation between comets, Hildas and quasi-Hildas could clarify the question. Toth (2006), makes an update of the inventory of this group. He found 11 new members of the quasi-Hilda cometary group of comets making a total of 15, and 23 outliers of the Hilda asteroid family in the Lagratian element space. He suspect that there could be some dormant or extinct comet nuclei among these asteroidal objects. If this were the case, the cometary nature of Hildas could be proved. We are going to attend the relation between Hildas and comets in section 4.

Evidence of impact cratering can be found in most of the solar system planets and satellites as well as on the population of the minor bodies. The collision among bodies was the fundamental mechanism during planetary accretion but also we have evidence of a present impact process at a slow rate. The collision of the Shoemaker-Levy 9 (SL9) comet with Jupiter in 1994 was a recent example of an impact between two celestial bodies in the present time. Craters are important markers in stablishing sequences of geologic events though also, impact structure forms may have changed through geologic time. So the study of the distribution of impact craters and their surrounding terrain is very important in order to establish an age sequence. Then, the study of the origin of the craters of a celestial body provide important information of both, the source and the target. As we mentioned, Hilda asteroids that escaped from the resonance are under the control of Jupiter, so we could expect there would be collisions between Hildas and the satellites of Jupiter. In particular we can study the Hilda cratering contribution to the four Galilean satellites: Io, Europa, Ganymede and Callisto. We are going to analyze this topic in section 5.

2. Hilda Asteroid Group

As we mentioned, in the 3:2 mean motion resonance with Jupiter it is present an important population, the Hilda asteroid group. They librate in a thin fringe in semimajor axis of 0.1 AU wide centered at ~ 3.97 AU. Up to the present there are 1744 Hildas cataloged in the ASTORB database.

They have a wide range of eccentricities but the great majority is clustered in the interval $0.1 < e < 0.3$, the inclinations extend up to $\sim 42°$. In Fig 2 we can see the distribution of semimajor axis and eccentricity of Hildas and their relative location to the Main Asteroid Belt.

This population is characterized by a central zone of a great stability (Nervorný et al. 1997, Ferraz Mello et al. 1998), but is surrounded by unstable regions. If an asteroid reaches these limit zone it can be quickly ejected from the resonance. Gil-Hutton and Brunini (2000) have shown that mutual collisions is the most efficient mechanism that, at present, is injecting Hilda asteroid fragments into the unstable regions beyond the boundaries of the resonance.

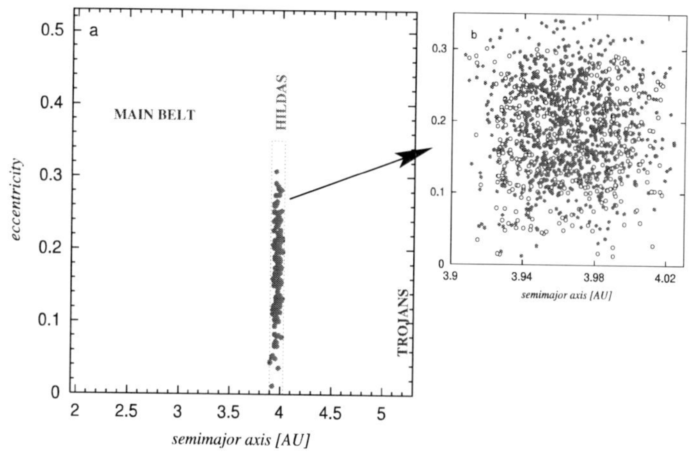

Figure 2. Asteroids in the a vs e plane and in detailed the Hilda zone where the red points are the real Hildas and the black empty circles are the fictitious Hildas generated in the simulation by Di Sisto et al. (2005).

The dynamical diffusion is very slow and then it should be taken into account the evaporation by mutual collisions to study the evolution of escaped Hildas (Brunini et al., 2003).

The biggest asteroid in the group is 153 Hilda with a diameter of 170.6 km. The whole population is characterized by a cumulative power low size distribution function in diameter (D), given by:

$$N(>D) \propto D^{2.11}, \tag{1}$$

The sample could be complete up to $D \sim 12$ km (Brunini et al. 2003). They are placed in the external zone of the asteroid belt, so they should have ices in their composition. Spectroscopic studies of Dahlgren and Lagerkvist (1995) and Dahlgren et al. (1997), reveal that Hilda asteroids are mainly of taxonomic classes D and P, which are associated to organic materials and ice enriched silicates. They also obtain a correlation between the taxonomic class and the size of Hildas. There are more D - type particles at smaller sizes.

However Hildas should have ices, they could be integrated or separated from the other material. If ices were integrated with the other material, sublimation could be more difficult and we could not detected it from Earth. But sublimation is also restricted by the formation of carbon mantles on the asteroid surface because their long exposure to the solar radiation in their stable resonance zone. So it is difficult from observations to really say the state of ices in Hilda asteroids. Only more dedicated observations could give light to this point.

3. Dynamical Evolution of Hilda Asteroids Removed from the Resonance

Di Sisto et al. (2005), studied the Hilda asteroid family like another probable source of JFCs. They performed a numerical integration of 500 fictitious Hilda asteroids under the gravitational influence of the Sun and the planets from Mercury to Neptune.

In Fig 2 we plot all asteroids in the semimajor axis vs eccentricity plane, and in detailed the Hilda zone, where the red points are the real Hildas and the black empty circles are our fictitious Hildas.

The sample was integrated for $1 \times 10^9 y$ which is an interval comparable to the age of the solar system. In this integration they followed the dynamical evolution of Hilda asteroids that escaped from the resonance. They obtained that 78.2% of them are ejected from the resonance and the rest, the 21.8% remains stable within it. In most of the cases, the asteroids left the resonance increasing first their eccentricity, and then their semimajor axis, in such a way that in the dynamical evolution they pass through the external solar system, having successive close encounters with the outer planets, mainly with Jupiter. The general dynamics of escaped Hildas is dominated by Jupiter's scattering. Therefore, at any time, most escaped Hildas may be found near jupiter' s orbit.

Being escaped Hildas under the gravitational control of Jupiter, one usual path of the dynamical evolution of them is the residence into a mean motion resonance with Jupiter (other than the 3:2) for some time. This behavior can be observed in the dynamical evolutions of three particles of the simulation by Di Sisto et al. (2005) shown in Fig. 3. In particular in Fig 2 b and c the particles go through the 1:1 resonance with Jupiter. The first particle transits that state from 3.1305×10^8 years to 3.1317×10^8 years, this is an interval of ~ 120000 years. The second one transits the 1:1 resonance from 590000 years to 685000 years, this is an interval of ~ 95000 years. Other frequent state of escaped particles is the pass for a Kozai resonance. A Kozai resonance occurs when the argument of pericentre, w, librates about a constant value. For low inclinations it is possible for w to librate about $w = 0°$ and $w = 180°$, and for large inclinations about $w = 90°$ and $w = 270°$. The semimajor axis of the object remains constant but the eccentricity and the inclination of the orbit are coupled in such a way that e is a maximum when i is a minimum, and vice versa.

4. Relation with JFCs

Comets and asteroids represent different parts of the planetesimal spectrum formed and processed in the Primitive Solar Nebula. The distinction between asteroids and comets is based in their observational qualities and orbital characteristics, rather than in their different properties or chemical composition. Comets are characterized by their *coma* of sublimated gas and dust, that gives them their typical aspect in the sky, but asteroids appear like light points in the sky.

Asteroids and JFC are indeed two dynamically different populations. Comets have highly eccentric orbits, being the great part of their orbits in the outer solar system, where their ices are preserved against depletion by solar heating. Asteroids have more circular orbits in proximity to the Sun and thus received much more solar heating over their lifetimes, making the survival of ices more difficult. The dynamical distinction is frequently defined by the Tisserand parameter with respect to Jupiter, T_j. The Tisserand parameter is an approximation of the Jacobi constant of motion in the circular restricted three body problem (CR3BP). In the problem with respect to Jupiter, the Tisserand parameter is given by

$$T_j = \frac{a_j}{a} + 2\sqrt{\frac{a}{a_j}\left(1 - e^2\right)}\ cos\, i, \tag{2}$$

where a, and e are the semimajor axis and eccentricity of the comet (or asteroid) and i is the inclination of the orbit with respect to the orbital plane of Jupiter.

The limit for the division between asteroids or comets is $T_j = 3$, since objects with $T_j < 3$ are dynamically coupled to Jupiter and then are considered cometary, while objects with $T_j > 3$ are decoupled to Jupiter and are considered asteroidal.

But although Hildas, in the external Main Belt, are in stable resonant motions and are protected from encountering Jupiter, once they escape from the resonance, they are dominated by the gravitation of Jupiter, and in this sense, they can resemble the dynamical state of a JFC.

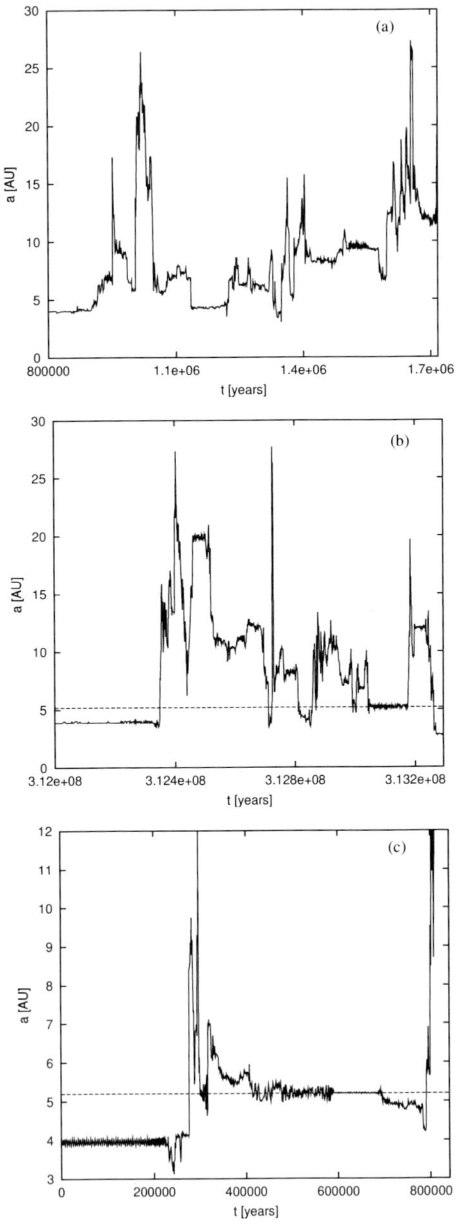

Figure 3. Dynamical evolution of three escaped Hildas. In fig 2 b and c the particles go through the 1:1 resonance with Jupiter at $a = 5.2$ AU for considerable intervals of time. This resonance is marked with the dashed line in both plots

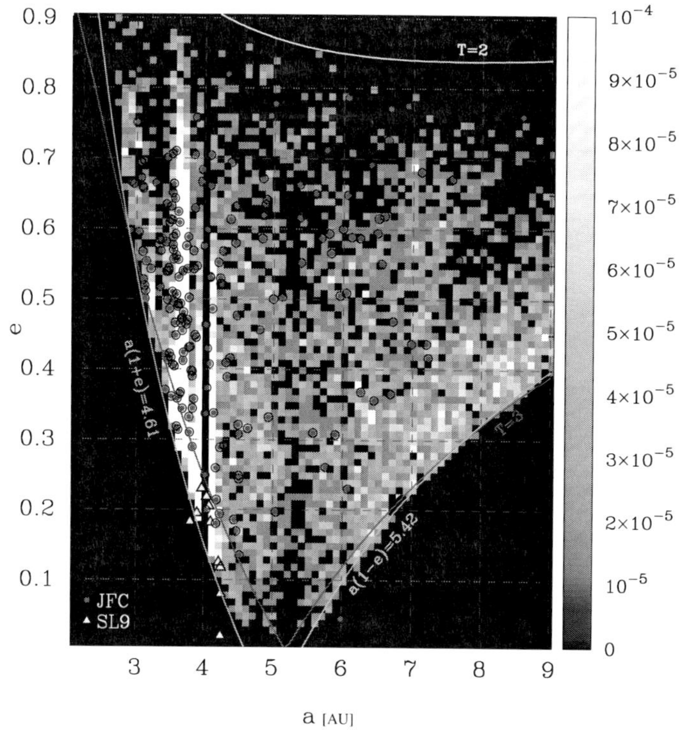

Figure 4. Time-weighted distribution of escaped Hildas in the a vs e space of the JFC zone. The red circles are the observed JFC. The yellow triangles are the pre-capture orbital elements of SL9 fragments (Chodas and Yeomans 1996)

Di Sisto et al. (2005) follow the evolution of escaped Hildas in order to determine if they pass some time like a JFC, assessing the population that, at present, contaminate the "genuine" population of JFC coming from the Trans Neptunian region. They define the JFC zone taking into account that JFCs pass through the Hill Sphere of Jupiter and have mostly semimajor axis between 2 and 9 AU and eccentricities less than 0.9. Besides, the comets are outside the 3:2 resonance. This criterion is almost equivalent to the range of Tisserand constant with respect to Jupiter $2 < T < 3$ proposed by Levison (1996) for the classification of JFCs. They obtained that from the bodies that left the resonance zone, 98.7% live at least 1000 years in the JFC region being their average residence time in this region of phase space 1.4×10^6 years. Also they calculated the number of escaped Hildas that are at present in the JFC region taking into account the colisional evaporation rate obtained in Brunini et al. (2003). Escaped Hildas mainly populate the JFC zone with $q > 2.5$ AU, and the contribution to the inner zone is negligible. It is expected that ~ 143 Hilda asteroids with diameter grater than 1 km are at present among the population of JFCs with $q > 2.5$ A.U., so they would be the $\sim 14\%$ of the JFC in that region. Although because physical life time of a comet is only 5 - 20 % of their dynamical lifetime, it could be that almost all the kilometer sized Hildas were dormant comets. They also estimate that the contribution of escaped Trojan asteroids to JFCs would be negligible.

Alvarez-Candal and Licandro (2006) studied the population of asteroids in cometary orbits (ACOs) in order to discriminate its cometary or asteroidal origin based on the comparison of cumulative size distributions of asteroids and comets. The ACOs are object with $T_j < 3$, so they would be cometary by the classification given above, but they do not present any signature of cometary activity. They divided the ACO population in two subsamples of ACOs in NEO (Near Earth Objects) orbits and non NEO orbits. They also analyzed the size distribution of Hildas and Trojans. They conclude that unstable Trojans and Hildas have size distribution indices similar to that of the NEO sample of ACOs, suggesting that a fraction could be scattered objects from the outer main belt. However Di Sisto et al. (2005) found that the contribution of escaped Hildas to the NEO population is negligible. On the other hand Alvarez-Candal and Licandro (2006) analyzed the size distribution index of JFCs and found that it is similar to that of the non-NEO sample of ACOs, suggesting that JFCs could be an important source of the non-NEO population.

Di Sisto et al. (2005) also give a picture of how escaped Hilda asteroids populate the JFC zone and so the Jupiter's neiborhood. In its Fig. 4, that is reproduced here in Figs. 4 and 5, they plot the distribution of residence times or time-weighted distribution of escaped Hildas obtained in the simulation in the orbital element space. They are also plotted the observed JFCs. As we see most of the JFCs are in the denser regions of escaped Hildas. The zone near the boundaries of the resonance is where escaped Hildas spend more time, then they increase their eccentricities leaving the stable region of the resonance. The break of the protection mechanism (libration around $\varphi \sim 0$) makes it possible close encounters with Jupiter. This causes the increase of their semimajor axis, crossing Jupiter's orbit and lefting quickly the JFC region. It is also seen a depletion of Hildas in the neighborhood of Jupiter's orbit at semimajor axis of $a_J \sim 5.2$ AU. A particle that crosses the Jupiter' s

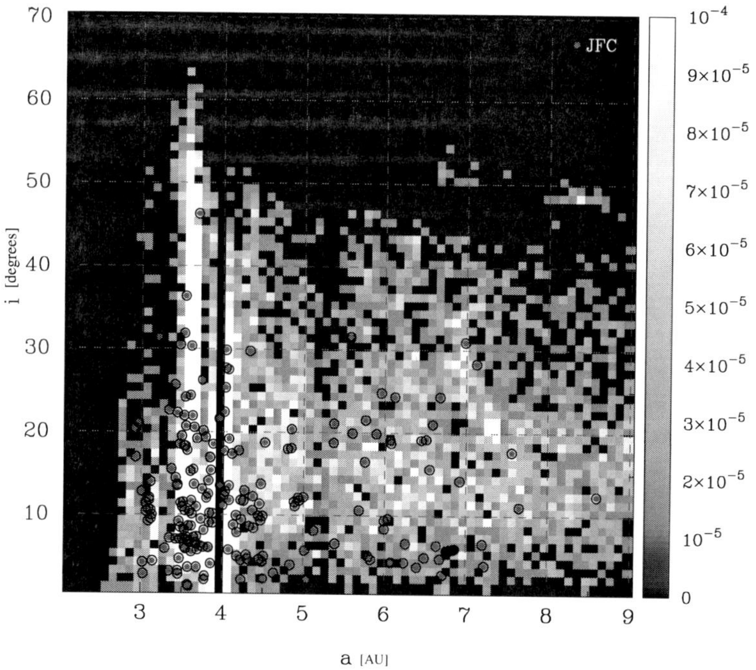

Figure 5. Time-weighted distribution of escaped Hildas in the a vs i space of the JFC zone. The red circles are the observed JFC.

orbit will have a close encounter with the planet that quickly removes it from the zone. This causes the depletion of asteroids with $a \sim a_J$.

4.1. The Relation with Quasi-Hildas and the SL9 Case

The study of the interrelation between the quasi Hilda cometary group of comets and Hilda asteroids could be broach from a dynamical point of view and also from the compositional point of view. The presence of comets inside or in the borders of the resonance is an evidence, both dynamical and physical, of the connection between them. As it is seen in Figs. 3 and 4 the probability of finding objects in the borders of the resonance is high and then is probable that comets observed in this zone be Hilda asteroids or fragments that have left the stable zone of the resonance. If they really were, the detection of activity would reveal the presence of ices in those asteroids and their comet nature. Although, the 3:2 resonance is a very stable zone, so Hildas have been there for a long time, and they could have slowly lost their surface ices by the continuous action of the Sun. Other phenomenon that is present is the formation of irradiation mantles on the asteroid surface. In any case the colisional activity produces fragments that could expose their trapped ices and then escape like potential comets. If we suppose that Hilda asteroids are really dormant comets, this population could be an important source for quasi-Hildas and also JFCs. They could be active or dormant, depending on their real composition and the time they have been exposed to the solar radiation.

A particular case is that of the Shoemaker-Levy 9 (SL9) comet that allow us to see in direct a collision of two celestial objects in the space. SL9 was tidally split into numerous fragments in 1992 after its perijove passage and then struck Jupiter in 1994. There was evidence of dust tails and out-gassing. Benner and Mc.Kinnon (1995), calculated the pre-capture orbital elements of SL9 and found two possible pre-capture orbits because of the chaoticity of the encounter. One of the possible pre-capture orbital element set overlaps the space of orbital elements of escaped Hildas and also the quasi-Hilda zone. This set of pre-capture orbital elements of SL9 is shown in Fig 4. So, at first, by virtue of this overlap we can say that an escaped Hilda could have been the parent of the fragments that struck Jupiter in 1994. Therefore, there is a chance that SL9 was an Hilda asteroid, and if it was, its pre-capture elements could be restricted.

5. Impact Cratering on the Galilean System

Impact cratering was a common process in the solar system as it is evident from the signatures left on the surfaces of most of its planets and satellites as well as on the population of the minor bodies. The collision among bodies was the fundamental mechanism during planetary accretion. Although the impact rate has been decreasing as the solar system was stabilizing, we have a considerable amount of evidence of an intense impact process in the past that continues at present at a much more slow rate. The determination of the present cratering rate on the surfaces of the solar system bodies furnishes one of the most fiducial ways of dating the geological time scale of activity.

The impact process of the inner solar system has been vastly studied, mainly, trough examinations of lunar cratering. To study the impact cratering history of bodies in the outer solar system the most natural and appropriate scenario are the four Galilean satellites: Io, Europa, Ganymede and Callisto.

Io has no known impact craters. Europa, is an icy world crossed by a network of dark fractures. The existence of few impact craters suggest that geologic processes are active today, and that it has a young surface. The largest Galilean satellite, the icy moon Ganymede, has a dark, heavily cratered terrain with more recent brighter grooved terrain. It also shows evidence of geologic activity, but its surface is very old, dating from the Late Heavy Bombardment or earlier. Also Callisto is very old. Its surface is completely saturated by craters. One feature unique to Callisto is the remnant structures of numerous impacts.

At looking for populations capable of producing craters on the Galilean surfaces we should investigate those objects that may be found at the distance of Jupiter. Zahnle et al. (1998), studied a number of populations that produced the craters on the Galilean satellites. They discussed in detail the Ecliptic Comets, using the numerical study of Levison and Duncan (1997). Other sources, like Long Period Comets, Trojan asteroids and asteroids from the main belt were also considered. Brunini et al. (2003) have investigated the possible contribution of escapees from the 3:2 mean motion resonance with Jupiter to the present cratering rate on the Galilean satellites. They shown that 8% of the objects leaving the 3:2 resonance impact Jupiter. Therefore, it would be natural that a fraction of them could also impact some of the Galilean Satellites. The impact rate on the surfaces of the four Galilean satellites was computed by Brunini et al. (2003) by assuming that after escaping from the resonance by collisional evolution, Hilda fragments follow a dynamical evolution similar to the one obtained in their numerical simulation of escaped Hildas (Di Sisto et al. 2005). From the 8 % of the fragments that end up impacting Jupiter, a small fraction (from 10^{-4} to 10^{-5}) impacts a given Galilean satellite. They do so at a velocity ranging from $\sim 15\ kms^{-1}$ for Callisto to $\sim 30\ kms^{-1}$ for Io. These impacts produce craters on the Galilean satellite surfaces whose diameter depend on several factors (e.g. the impact velocity, the mass of the target and the projectile mass).

The mean interval between the production of craters of two different diameter D computed by Brunini et al. (2003) are shown in Table 1.

The other main source of craters are the Ecliptic comets. Zahnle et al. (1998) computed the mean intervals between craters produced by them. They are 2.6×10^7 years on Io, 1.5×10^7 years on Europe, 1.0×10^7 years on Ganymede and 2.6×10^7 years on Callisto. These timescales are comparable to the time scales for the production of craters of $D > 10km$ by Hilda asteroids,

Table 1. Mean interval between impacts on the Galilean satellites in years, capable to produce craters with $D > 10km$ km and $D > 4km$. For these calculations the radius of the minimum target considered to be member of the Hilda population was $0.4km$.

Cratering rate

Satellite	$\Delta T(D = 10km)$	$\Delta T(D = 4km)$
Io	$2.1 \times 10^7 y$	$2.0 \times 10^6 y$
Europe	$1.5 \times 10^7 y$	$1.8 \times 10^6 y$
Ganymede	$1.1 \times 10^7 y$	$1.3 \times 10^6 y$
Callisto	$2.6 \times 10^7 y$	$3.3 \times 10^6 y$

but are longer than the time scales for $D > 4km$. Therefore, Brunini et al. (2003) conclude that Hilda asteroids is the main source of small craters on the surfaces of the Galilean satellites.

Europa has only 27 known craters with $D > 4km$ (Moore et al. 2001). Although Bierhaus et al. 2001 states that the vast majority of small craters are secondaries, probably all craters larger than 2 km are primary craters. So, $D > 4km$ seems to be a safe election for primary craters.

One difference between the cratering rates from Hilda asteroids and Ecliptic comets is regarding the time-scales involved in the process. Brunini et al. (2003) shown that nearly 75% of the objects that impact Jupiter do so in less than 5×10^4 yr after escape from the resonance. It represents a very short time scale compared with the 72 My that typically spends a comet to travel from the Kuiper Belt to the Jupiter region (Di Sisto and Brunini 2007). We could speculate that after a catastrophic collision, many fragments escape from the Hilda region. A considerable number of them end-up colliding Jupiter (~ 8 %) during the relatively short time interval of some 10^4 yr. If some signature of this kind of process is found on the surfaces of the Gallilean satellites, it would furnish important clues regarding the past history of the Hildas.

6. Conclusion

In this paper we have review the contribution of Hilda asteroids to the Jupiter neiborhood. We have seen how escaped Hildas, recently removed from their stable orbits populate the zone of JFCs and its contribution to the cratering rate on the major satellites of Jupiter. We can point the following items:

- Hilda asteroids contribute with $\sim 14\%$ to the JFC population with $q > 2.5$ A.U.. Almost all the escaped Hildas pass through the dynamical region occupied by JFCs, and the mean dynamical life time there is 1.4×10^6 years. The more abundant taxonomic class in the Hilda group is D, with more D asteroids with less diameter. Then, as escaped Hildas are fragments resulting from a catastrophic collision, it is more probable that the escaped objects were D types. This is the more primitive class and the more similar to comets.

- Escaped Hildas spend more time in the zone near the boundaries of the resonance. If Hilda asteroids were really dormant comets, this population could be an important source for quasi-Hildas.

- One of the two sets of pre-capture orbital elements of SL9 obtain by Benner and Mc.Kinnon (1995) overlaps with the orbital elements of the Hildas in the JFC zone, supporting the idea that it would be an Hilda asteroid, and if it was, the pre-capture orbital elements would be defined.

- From numerical simulations of the dynamical evolution of fictitious Hilda asteroids, 8 % of the particles that leave the resonance hit Jupiter. Hilda asteroid population could be the main source of small impact craters on Jupiter satellites, overcoming the contribution of ecliptic comets calculated by Zahnle et al. (1998).

- There are some open issues in relation to the escaped Hildas that is convenient to point out. The existence of the quasi-Hilda comets in the borders of the 3:2 resonance could give the keys of the relation between asteroids and comets. It is needed more observations of Hilda asteroids directed to detect activity and study their surface compositions in order to understand the real nature of Hildas. On the other hand, there is a short time scale involved in the production of craters on the Galilean satellites from escaped Hildas. So this process must leave some signature, for example sets of small craters on the surfaces of the satellites produced by many escaped fragments of a catastrophic collision in the Hilda zone. These characteristics could be searched by dedicated spatial missions.

References

[1] Alvarez-Candal, A. and Licandro, J. The size distribution of asteroids in cometary orbits and related populations. *A&A* **458**, Issue 3, 1007-1011 (2006).

[2] Benner, L.A.M., Mc.Kinnon, W.B. On the orbital evolution and origin of Comet Shoemaker-Levy 9. *Icarus*. **118**, 155-168 (1995).

[3] Bierhaus, E. B., C. R. Chapman, W. J. Merline, S. M. Brooks and E. Asphaug. Pwyll Secondaries and Other Small Craters on Europa. *Icarus*. **153**, 264-276 (2001).

[4] Brunini, A., Di Sisto R.P., and Orellana R.B. Cratering rate on the jovian system: the contribution from Hilda asteroids. *Icarus*. **165**, 371-378 (2003).

[5] Dahlgren, M. and C.-I. Lagerkvist. A study of Hildas asteroids. I. CCD spectroscopy of Hilda asteroids. *Astronomy and Astrophysics*. **302**, 907-914 (1995).

[6] Dahlgren, M., C.-I. Lagerkvist, A. Fitzsimmons, I.P. Williams. and M. Gordon. A study of Hildas asteroids. II. Compositional implications from optical spectroscopy. *Astronomy and Astrophysics*. **323**, 606-619 (1997).

[7] Chodas, P., W. & Yeomans, D., K. *IAU Colloq. 156: The Collision of Comet Shoemaker-Levy 9 and Jupiter* (eds Noll, K. S., Feldman, P. D. & Weaver, H. A.), 1-30 (1996).

[8] Di Sisto R. P., A. Brunini, L. D. Dirani and R. B. Orellana. Hilda asteroids among Jupiter family comets. *Icarus* **174**, 81-89 (2005).

[9] Di Sisto R. P. and A. Brunini. The origin and distribution of the Centaur population *Icarus* **190**, Issue 1, 224-235 (2007).

[10] Fernández, J. A. On the existence of a comet belt beyond Neptune. *Mon. Not. R. Astron. Soc.* **192**,481-491 (1980).

[11] Ferraz-Mello, S., Michtchenko, T. A. Nesvorný, D., Roig, F. and Simula, A. The depletion of the Hecuba gap vs. the long-lasting Hilda group. *Planet. Space Sci.*. **46**, No 11/12 1425-1432 (1998).

[12] Gil-Hutton, R. & Brunini, A. Collisional Evolution of the Outer Asteroid Belt. *Icarus* **145**, 382-390 (2000).

[13] Hartmann, W.K., D.J. Tholen and D.P.Cruikshank. The Relationship of active Comets, "Extinct" Comets, and Dark Asteroids *Icarus* **69**, 33-50 (1987).

[14] Kresák, L. Dynamical interrelations among comets an asteroids. In Asteroaids (T. Gehrels, Ed.) pp. 289-309. Univ. of Arizona Press, Tucson (1979).

[15] Levison, H. Comet taxonomy. In *Completing the Inventory of the Solar System* (T.W. Retting and J. M. Hahn, Eds.), Astron. Soc. of Pacific Conference Proceedings, **107**, 233-244 (1996).

[16] Levison, H. and M. Duncan. From the Kuiper Belt to Jupiter-family comets: The spatial distribution of ecliptic comets. *Icarus* **127**, 13-32 (1997).

[17] Moore, J. M., and 25 collegues. Impact features on Europa: Results of the Galileo Europa Mission. *Icarus* **151**, 93-111 (2001).

[18] Nervorný, D. and Ferraz Mello, S. On the Asteroidal Population of the First-Order Jovian Resonances. *Icarus* **130**, 247-258. (1997).

[19] Toth, i. The quasi-Hilda subgroup of ecliptic comets - an update (Research Note) *A&A* **448**, 1191-1196 (2006).

[20] Zahnle, K., Dones, L. and H.F. Levison. Cratering Rates on the Galilean Satellites. *Icarus*. **136**, 202-222 (1998).

In: Meteorites and Asteroids
Editors: A. Dementieva et al.
ISBN: 978-1-61942-115-8

Chapter 7

ASTEROIDS APOPHIS AND 1950 DA: 1000 YEARS ORBIT EVOLUTION AND POSSIBLE USE

Joseph J. Smulsky[a,*] ***and Yaroslav J. Smulsky***[b]
[a] Institute of the Earth's Cryosphere, Siberian Branch of Russian Academy of Sciences, Tyumen, Russia
[b] Institute of Thermophysics of Russian Academy of Sciences, Siberian Branch, Novosibirsk, Russia

ABSTRACT

The evolution of movement and possible use two asteroids is examined: Apophis and 1950 DA. As a result of the analysis of publications it is established that uncertainty of trajectories of Apophis are caused by imperfection of methods of its determination. The differential equations of motion of Apophis, planets, the Moon and the Sun are integrated by new numerical method and the evolution of the asteroid orbit is investigated. The Apophis will pass by the Earth at a distance of 6.1 its radii on April 13th, 2029. It will be its closest approach with the Earth during next 1000 years. A possibility of transformation of Apophis orbit to an orbit of the Earth's satellite, which can be used for various tasks, is considered. The similar researches have been executed for asteroid 1950 DA. The asteroid will twice approach the Earth to a minimal distance of 2.25 million km, in 2641 and in 2962. It can be made an Earth-bound satellite by increasing its aphelion velocity by ~ 1 km s^{-1} and by decreasing its perihelion velocity by ~ 2.5 km s^{-1}.

Keywords: Near-Earth Objects; Asteroids, dynamics; Satellites, dynamics

1. Introduction

Over the past decade, the asteroids of prime interest have been two asteroids, Apophis and 1950 DA, the first predicted to approach the Earth in 2029, and the second, in 2880. Reported calculations revealed some probability of an impact of the asteroids on the Earth. Yet, by the end of the decade refined orbital-element values of the asteroids were obtained, and more precise algorithms for calculating the interactions among solar-system bodies were developed. Following this, in the present paper we consider the motion evolution of both asteroids. In addition, we discuss available possibilities for making the asteroids into the Earth-bound satellites. Initially, the analysis is applied to Apophis and, then, numerical data for 1950 DA obtained by the same method will be presented.

The background behind the problem we treat in the present study was recently outlined in Giorgini *et al.* 2008. On June 19 – 20, 2004, asteroid Apophis was discovered by astronomers at the Kitt Peak Observatory (Tucker *et al.* 2004), and on December 20, 2004 this asteroid was observed for the second time by astronomers from the Siding Spring Survey Observatory (Garradd 2004). Since then, the new asteroid has command international attention. First gained data on identification of Apophis' orbital elements were employed to predict the Apophis path. Following the first estimates, it was reported in Rykhlova *et al.* 2007 that on April 13, 2029 Apophis will approach the Earth center to a minimum distance of 38000 km. As a result of the Earth gravity, the Apophis orbit will alter appreciably. Unfortunately, presently available methods for predicting the travel path of extraterrestrial objects lack sufficient accuracy, and some authors have therefore delivered an opinion that the Apophis trajectory will for long remain unknown, indeterministic, and even chaotic (see Giorgini *et al.* 2008, Rykhlova *et al.* 2007, Emel'yanov *et al.* 2008a). Different statistical predictions points to some probability of Apophis' collision with the Earth on April 13, 2036. It is this aspect, the impact risk, which has attracted primary attention of workers dealing with the problem.

Rykhlova *et al.* 2007 have attempted an investigation into the possibility of an event that the Apophis will closely approach the Earth. They also tried to evaluate possible threats stemming from this event. Various means to resist the fall of the asteroid onto Earth were put forward, and proposals for tracking Apophis missions, made. Finally, the need for prognostication studies of the Apophis path accurate to a one-kilometer distance for a period till 2029 was pointed out.

Many points concerning the prospects for tracking the Apophis motion with ground- and space-based observing means were discussed in Giorgini *et al.* 2008, Rykhlova *et al.* 2007, Emel'yanov *et al.* 2008a, 2008b. Since the orbits of the asteroid and Earth pass close to each other, then over a considerable portion of the Apophis orbit the asteroid disc will only be partially shined or even hidden from view. That is why it seems highly desirable to identify those periods during which the asteroid will appear accessible for observations with ground means. In using space-based observation means, a most efficient orbital allocation of such means needs to be identified.

Prediction of an asteroid motion presents a most challenging problem in astrophysics. In Sokolov *et al.* 2008, the differential equations for the perturbed motion of the asteroid were integrated by the Everhart method (Everhart 1974); in those calculations, for the coordinates of perturbing bodies were used the JPL planetary ephemeris DE403 and DE405 issued by the Jet Propulsion Laboratory, USA. Sufficient attention was paid to resonance phenomena that might cause the hypothetical 2036 Earth impact.

Bykova and Galushina 2008*a*, 2008*b* used 933 observations to improve the identification accuracy for initial Apophis orbital parameters. Yet, the routine analysis has showed that, as a result of the pass of the asteroid through several resonances with Earth and Mars, the motion of the asteroid will probably become chaotic. With the aim to evaluate the probability of an event that Apophis will impact the Earth in 2036, Bykova *et al.* 2008 have made about 10 thousand variations of initial conditions, 13 of which proved to inflict a fall of Apophis onto Earth.

Smirnov 2008 has attempted a test of various integration methods for evaluating their capabilities in predicting the motion of an asteroid that might impact the Earth. The Everhart method, the Runge-Kutta method of fourth order, the Yoshida methods of sixth and eighth orders, the Hermit method of fourth and sixth orders, the Multistep Predictor-Corrector (MS-PC) method of sixth and eighth orders, and the Parker-Sochacki method were analyzed. The Everhart and MS-PC methods proved to be less appropriate than the other methods. For example, at close Apophis-to-Earth distances E.A. Smirnov used, instead of the Everhart method, the Runge-Kutta method. He to the fact that, in the problems with singular points, finite-difference methods normally fail to accurately approximate higher-order derivatives. This conclusion is quite significant since below we will report on an integration method for motion equations free of such deficiencies.

In Ivashkin and Stikhno 2008 the mathematical problems on asteroid orbit prediction and modification were considered. Possibilities offered by the impact-kinetic and thermonuclear methods in correcting the Apophis trajectory were evaluated.

An in-depth study of the asteroid was reported in Giorgini *et al.* 2008. A chronologically arranged outline of observational history was given, and the trend with progressively reduced uncertainty region for Apophis' orbit-element values was traced. Much attention was paid to discussing the orbit prediction accuracy and the bias of various factors affecting this accuracy. The influence of uncertainty in planet coordinates and in the physical characteristics of the asteroid, and also the perturbing action of other asteroids, were analyzed. The effects on integration accuracy of digital length, non-spherical shape of Earth and Moon, solar-radiation-induced perturbations, non-uniform thermal heating, and other factors, were examined.

The equations of perturbed motion of the asteroid were integrated with the help of the Standard Dynamic Model (SDM), with the coordinates of other bodies taken from the JPL planetary ephemeris DE405. It is a well-known fact that the DE405 ephemerid was compiled as an approximation to some hundred thousand observations that were made till 1998. Following the passage to the ephemeris DE414, that approximates observational data till 2006, the error in predicting the Apophis trajectory on 2036 has decreased by 140000 km. According to Giorgini *et al.* 2008, this error proved to be ten times greater than the errors induced by minor perturbations. Note that this result points to the necessity of employing a more accurate method for predicting the asteroid path.

In Giorgini *et al.* 2008, prospects for further refinement of Apophis' trajectory were discussed at length. Time periods suitable for optical and radar measurements, and also observational programs for oppositions with Earth in 2021 and 2029 and spacecraft missions for 2018 and 2027 were scheduled. Future advances in error minimization for asteroid trajectory due to the above activities were evaluated.

It should be noted that the ephemerides generated as an approximation to observational data enable rather accurate determination of a body's coordinates in space within the approximation time interval. The prediction accuracy for the coordinates on a moment remote from this interval worsens, the worsening being the greater the more the moment is distant from the approximation interval. Therefore, the observations and the missions scheduled in Giorgini *et al.* 2008 will be used in refining future ephemerides.

In view of the afore-said, in calculating the Apophis trajectory the equation of perturbed motion were integrated (Giorgini *et al.* 2008, Sokolov *et al.* 2008, Ivashkin and Stikhno 2008), while the coordinates of other bodies were borrowed from the ephemerid. Difference integration methods were employed, which for closely spaced bodies yield considerable inaccuracies in calculating higher-order derivatives. Addition of minor interactions to the basic Newtonian gravitational action complicates the problem and enlarges the uncertainty region in predicting the asteroid trajectory. Many of the weak interactions lack sufficient quantitative substantiation. Moreover, the physical characteristics of the asteroid and the interaction constants are known to some accuracy. That is why in making allowance for minor interactions expert judgments were used. And, which is most significant, the error in solving the problem on asteroid motion with Newtonian interaction is several orders greater than the corrections due to weak additional interactions.

The researches, for example, Bykova and Galushina 2008*a*, 2008*b* apply a technique in Giorgini *et al.* 2008 to study of influence of the initial conditions on probability of collision Apophis with Earth. The initial conditions for asteroid are defined from elements of its orbit, which are known with some uncertainty. For example, eccentricity value $e=e_n\pm\sigma_e$, where e_n is nominal value of eccentricity, and σ_e is root-mean-square deviation at processing of several hundred observation of asteroid. The collision parameters are searched in the field of possible motions of asteroid, for example for eccentricitya, $3\sigma_e$, the initial conditions are calculated in area $e=e_n\pm\sigma_e$. From this area the 10 thousand, and in some works, the 100 thousand sets of the initial conditions are chosen by an accidental manner, i.e. instead of one asteroid it is considered movement 10 or 100 thousand asteroids. Some of them can come in collision with Earth. The probability of collision asteroid with the Earth is defined by their amount.

Such statistical direction is incorrect. If many measurement data for a parameter are available, then the nominal value of the parameter, say, eccentricity e_n, presents a most reliable value for it. That is why a trajectory calculated from nominal initial conditions can be regarded as a most reliable trajectory. A trajectory calculated with a small deviation from the nominal initial conditions is a less probable trajectory, whereas the probability of a trajectory calculated from the parameters taken at the boundary of the probability region (i.e. from $e = e_n \pm \sigma_e$) tends to zero. Next, a trajectory with initial conditions determined using parameter values trice greater than the probable deviations (i.e. $e = e_n \pm 3\sigma_e$) has an even lower, negative, probability. Since initial conditions are defined by six orbital elements, then simultaneous realization of extreme (boundary) values ($\pm\ 3\sigma$) for all elements is even a less probable event, i.e. the probability becomes of smaller zero.

That is why it seems that a reasonable strategy could consist in examining the effect due to initial conditions using such datasets that were obtained as a result of successive accumulation of observation data. Provided that the difference between the asteroid motions in the last two datasets is insignificant over some interval before some date, it can be concluded that until this date the asteroid motion with the initial conditions was determined quite reliably.

As it was shown in Giorgini *et al.* 2008, some additional activities are required, aimed at further refinement of Apophis' trajectory. In this connection, more accurate determination of Apophis' trajectory is of obvious interest since, following such a determination, the range of possible alternatives would diminish.

For integration of differential motion equations of solar-system bodies over an extended time interval, a program Galactica was developed (Grebenikov and Smulsky 2007, Melnikov and Smulsky 2009). In this program, only the Newtonian gravity force was taken into account, and no differences for calculating derivatives were used. In the problems for the compound model of Earth rotation (Mel'nikov *et al.* 2008) and for the gravity maneuver near Venus (Smulsky 2008), motion equations with small body-to-body distances, the order of planet radius, were integrated. Following the solution of those problems and subsequent numerous checks of numerical data, we have established that, with the program Galactica, we were able to rather accurately predict the Apophis motion over its travel path prior to and after the approach to the Earth. In view of this, in the present study we have attempted an investigation into orbit evolution of asteroids Apophis and 1950 DA; as a result of this investigation, some fresh prospects toward possible use of these asteroids have opened.

2. Problem Statement

For the asteroid, the Sun, the planets, and the Moon, all interacting with one another by the Newton law of gravity, the differential motion equations have the form (Smulsky 1999):

$$\frac{d^2\vec{r}_i}{dt^2} = -G\sum_{k\neq i}^{n}\frac{m_k\vec{r}_{ik}}{r_{ik}^{3}},\ i = 1,2,\ldots, n, \tag{1}$$

where $\vec{r}_i$ is radius-vector of a body with mass m_i relatively Solar System barycenter; G is gravitational constant; $\vec{r}_{ik}$ is vector $\vec{r}_i - \vec{r}_k$ and r_{ik} is its module; $n = 12$.

As a result of numerical experiments and their analysis we came to a conclusion, that finite-difference methods of integration do not provide necessary accuracy. For the integration of Eq. (1) we have developed algorithm and program Galactica. The meaning of function at the following moment of time $t=t_0 + \Delta t$ is determined with the help of Taylor series, which, for example, for coordinate x looks like:

$$x = x_0 + \sum_{k=1}^{K}\frac{1}{k!}x_0^{(k)}(\Delta t)^k, \tag{2}$$

where $x_0^{(k)}$ is derivative of k order at the initial moment t_0.

The meaning of velocity x' is defined by the similar formula, and acceleration x_0'' by the Eq. (1). Higher derivatives $x_0^{(k)}$ are determined analytically as a result of differentiation of the Eq. (1). The calculation algorithm of the sixth order is now used, i.e. with K=6.

3. Preparation of Initial Data

We consider the problem of interest in the barycentric coordinate system on epoch J2000.0, Julian day JD_s = 2451545. The orbital elements asteroids Apophis and 1950 DA, such as the eccentricity e, the semi-major axis a, the ecliptic obliquity i_e, the ascending node angle Ω, the ascending node-perihelion angle ω_e, etc., and asteroids position elements, such as the mean anomaly M, were borrowed from the JPL Small-Body database 2008 as specified on November 30.0, 2008. The data, represented to 16 decimal digits, are given in Table 1. For Apophis in Table 1 the three variants are given. The first variant is now considered. These elements correspond to the solution with number JPL sol. 140, which is received Otto Mattic at April 4, 2008. In Table 1 the uncertainties of these data are too given. The relative uncertainty value δ

is in the range from $2.4{\cdot}10^{-8}$ to $8{\cdot}10^{-7}$. The same data are in the asteroid database by Edward Bowell 2008, although these data are represented only to 8 decimal digits, and they differ from the former data in the 7-th digit, i.e., within value δ. Giorgini *et al.* 2008 used the orbital elements of Apophis on epoch JD = 2453979.5 (September 01.0, 2006), which correspond to the solution JPL sol. 142. On publicly accessible JPL-system Horizons the solution sol. 142 can be prolonged till November 30.0, 2008. In this case it is seen, that difference of orbital elements of the solution 142 from the solution 140 does not exceed 0.5σ uncertainties of the orbit elements.

The element values in Table 1 were used to calculate the Cartesian coordinates of Apophis and the Apophis velocity in the barycentric equatorial system by the following algorithm (see Duboshin 1976, Smulsky 2007, Mel'nikov *et al.* 2008, Melnikov and Smulsky 2009).

From the Kepler equation

$$E - e{\cdot}\sin E = M, \tag{3}$$

we calculate the eccentric anomaly E and, then, from E, the true anomaly φ_0:

$$\varphi_0 = 2 \cdot \mathrm{arctg}[\sqrt{(1+e)/(1-e)} \cdot \mathrm{tg}(0.5 \cdot E)], \tag{4}$$

In subsequent calculations, we used results for the two-body interaction (the Sun and the asteroid) (Smulsky 2007, Smulsky 2008). The trajectory equation of the body in a polar coordinate system with origin at the Sun has the form:

$$r = \frac{R_p}{(\alpha_1 + 1)\cos\varphi - \alpha_1}, \tag{5}$$

where the polar angle φ, or, in astronomy, the true anomaly, is reckoned from the perihelion position $r = R_p$; $\alpha_1 = -1/(1+e)$ is the trajectory parameter;

Table 1. Three variants of orbital elements of asteroids Apophis on two epochs and 1950 DA on one epoch in the heliocentric ecliptic coordinate system of 2000.0 with $JD_S = 2451545$ (see JPL Small-Body Database 2008)

Elements	Apophis				1950 DA	Units
	1-st variant November 30.0, 2008 $JD_{01} = 2454800.5$ JPL sol.140	Uncertainties $\pm\sigma$ 1-st var.	2-nd variant January 04.0 2010 $JD_{02} = 2455200.5$ JPL sol.144	3-rd variant November 30.0, 2008 $JD_{01} = 2454800.5$ JPL sol.144.	November 30.0, 2008 $JD_0 = 2454800.5$ JPL sol. 51	
	Magnitude					
e	.1912119299890948	7.6088e-08	.1912110604804485	.1912119566344382	0.507531465407232	
a	.9224221637574083	2.3583e-08	.9224192977379344	.9224221602386669	1.698749639795436	AU
q	.7460440415606373	8.6487e-08	.7460425256098334	.7460440141364661	0.836580745750051	AU
i_e	3.331425002325445	2.024e-06	3.331517779979046	3.331430909298658	12.18197361251942	deg
Ω	204.4451349657969	0.00010721	204.4393039605681	204.4453098275707	356.782588306221	deg
ω_e	126.4064496795719	0.00010632	126.4244705298442	126.4062862564680	224.5335527346193	deg
M	254.9635275775066	5.7035e-05	339.9486156711335	254.9635223452623	161.0594270670401	deg
t_p	2454894.912750123770 (2009-Mar- 04.41275013)	5.4824e-05	2455218.523239657948 (2010-Jan-22.02323966)	2454894.912754286546 (2009-Mar-04. 41275429)	2.454438.693685309 (2007-Dec- 12.0419368531	JD d
P	323.5884570441701 0.89	1.2409e-05 3.397e-08	323.5869489330219 0.89	323.5884551925927 0.89	808.7094041052905 2.21	D yr
n	1.112524233059586	4.2665e-08	1.112529418096263	1.112524239425464	0.445153720449539	deg/d
Q	1.098800285954179	2.8092e-08	1.098796069866035	1.098800306340868	2.560918533840822	AU

Table 2. The masses m_{bj} of the planets from Mercury to Pluto, the Moon, the Sun (1 – 11) and asteroids: Apophis (12a) and 1950 DA (12b), and the initial condition on epoch JD_0 = 2454800.5 (November 30.0, 2008) in the heliocentric equatorial coordinate system on epoch 2000.0 JD_S = 2451545. G = 6.67259E-11 $m^3 s^{-2} \cdot kg^{-1}$

Bodies, j	Bodies masses in *kg*, their coordinates in *m* and velocities in $m\ s^{-1}$			
	m_{bj}	x_{aj}, v_{xaj},	y_{aj}, v_{yaj}	z_{aj}, v_{zaj}
1	3.30187842779737E+23	-17405931955.9539	-60363374194.7243	-30439758390.4783
		37391.7107852059	-7234.98671125365	-7741.83625612424
2	4.86855338156022E+24	108403264168.357	-2376790191.8979	-7929035215.64079
		1566.99276862423	31791.7241663148	14204.3084779893
3	5.97369899544255E+24	55202505242.89	125531983622.895	54422116239.8628
		-28122.5041342966	10123.4145376039	4387.99294255716
4	6.4185444055007E+23	-73610014623.8562	-193252991786.298	-86651102485.4373
		23801.7499674501	-5108.24106287744	-2985.97021694235
5	1.89900429500553E+27	377656482631.376	-609966433011.489	-270644689692.231
		11218.8059775149	6590.8440254003	2551.89467211952
6	5.68604198798257E+26	-1350347198932.98	317157114908.705	189132963561.519
		-3037.18405985381	-8681.05223681593	-3454.56564456648
7	8.68410787490547E+25	2972478173505.71	-397521136876.741	-216133653111.407
		979.784896813787	5886.28982058747	2564.10192504801
8	1.02456980223201E+26	3605461581823.41	-2448747002812.46	-1092050644334.28
		3217.00932811768	4100.99137103454	1598.60907148943
9	1.65085753263927E+22	53511484421.7929	-4502082550790.57	-1421068197167.72
		5543.83894965145	-290.586427181992	-1757.70127979299
10	7.34767263035645E+22	55223150629.6233	125168933272.726	54240546975.7587
		-27156.1163326908	10140.7572420768	4468.97456956941
11	1.98891948976803E+30	0	0	0
		0	0	0
12a	30917984100.3039	-133726467471.667	-60670683449.3631	-26002486763.62
		16908.9331065445	-21759.6060221801	-7660.90393288287
12b	1570796326794.9	314388505090.346	171358408804.935	127272183810.191
		-5995.33838888362	9672.35319009371	6838.06006342785

and $R_p = a \cdot (2\alpha_1 + 1)/\alpha_1$ is the perihelion radius. The expressions for the radial υ_r and transversal υ_t velocities are

$$\upsilon_r = \upsilon_p \sqrt{(\alpha_1 + 1)^2 - (\alpha_1 + 1/\bar{r})^2}, \quad \text{for} \quad \varphi\ \varphi > \pi \quad \text{we have } \upsilon_r < 0;$$

$$\upsilon_t = \upsilon_p / \bar{r}, \tag{6}$$

where $\bar{r} = r/R_p$ is the dimensionless radius, and the velocity at perihelion is

$$\upsilon_p = \sqrt{G(m_S + m_{As})/(-\alpha_1)R_p)}\,, \tag{7}$$

where $m_S = m_{11}$ is the Sun mass (the value of m_{11} is given in Table 2), and $m_{As}=m_{12}$ is the Apophis mass.

The time during which the body moves along an elliptic orbit from the point of perihelion to an orbital position with radius $\bar{r}$ is given by

$$t = \frac{R_p}{\upsilon_p}\left[\frac{\bar{r}\,|\,\bar{\upsilon}_r\,|}{2\alpha_1+1} - \frac{\alpha_1\left(\pi/2 + \arcsin\{[(2\alpha_1+1)\bar{r} - \alpha_1]/(-\alpha_1-1)\}\right)}{(-2\alpha_1-1)^{3/2}}\right], \tag{8}$$

where $\bar{\upsilon}_r = \upsilon_r/\upsilon_p$ is the dimensionless radial velocity.

At the initial time $t_0 = 0$, which corresponds to epoch JD_0 (see Table 1), the polar radius of the asteroid r_0 as dependent on the initial polar angle, or the true anomaly φ, can be calculated by Eq. (5)The initial radial and initial transversal velocities as functions of r_0 can be found using Eq. (6).

The Cartesian coordinates and velocities in the orbit plane of the asteroid (the axis x_o goes through the perihelion) can be calculated by the formulas

$$x_o = r_0 \cdot \cos\varphi_0;\ y_o = r_0 \cdot \sin\varphi_0; \tag{9}$$

$$\upsilon_{xo} = \upsilon_r \cdot \cos\varphi_0 - \upsilon_t \cdot \sin\varphi_0\,;\ \upsilon_{yo} = \upsilon_r \cdot \sin\varphi_0 + \upsilon_t \cdot \cos\varphi_0\,. \tag{10}$$

The coordinates of the asteroid in the heliocentric ecliptic coordinate system can be calculated as

$$x_e = x_o \cdot (\cos\omega_e \cdot \cos\Omega - \sin\omega_e \cdot \sin\Omega \cdot \cos i_e) - y_o \cdot (\sin\omega_e \cdot \cos\Omega + \cos\omega_e \cdot \sin\Omega \cdot \cos i_e); \tag{11}$$

$$y_e = x_o \cdot (\cos\omega_e \cdot \sin\Omega - \sin\omega_e \cdot \cos\Omega \cdot \cos i_e) - y_o \cdot (\sin\omega_e \cdot \sin\Omega - \cos\omega_e \cdot \cos\Omega \cdot \cos i_e); \tag{12}$$

$$z_e = x_o \sin\omega_e \cdot \sin i_e + y_o \cdot \cos\omega_e \cdot \sin i_e. \tag{13}$$

The velocity components of the asteroid $\upsilon_{xe}, \upsilon_{ye}$ and υ_{ze} in this coordinate system can be calculated by Equations analogous to (11) – (13).

Since Eq. (1) are considered in a motionless equatorial coordinate system, then elliptic coordinates (11) – (13) can be transformed into equatorial ones by the Equations

$$x_a = x_e;\ y_a = y_e \cdot \cos \varepsilon_0 \cdot - z_e \cdot \sin \varepsilon_0;\ z_a = y_e \cdot \sin \varepsilon_0 + z_e \cdot \sin \varepsilon_0, \qquad (14)$$

where $\varepsilon_0 \cdot$ is the angle between the ecliptic and the equator in epoch JD_S.

The velocity components $\upsilon_{xe}, \upsilon_{ye}$ and υ_{ze} can be transformed into the equatorial ones $\upsilon_{xa}, \upsilon_{ya}$ and υ_{za} by Equations analogous to (14). With known heliocentric equatorial coordinates of the Solar system n bodies x_{ai}, y_{ai}, z_{ai} i = 1,2, … n, the coordinates of Solar system barycentre, for example, along axis x will be:

$X_c = (\sum_{i=1}^{n} m_i x_{ai}) / M_{Ss}$, where $M_{Ss} = \sum_{i=1}^{n} m_i$ is mass of Solar system bodies.

Then barycentric equatorial coordinates x_i of asteroid and other bodies will be

$$x_i = x_{ai} - X_c.$$

Other coordinates y_i and z_i and components of velocity $\upsilon_{xi}, \upsilon_{yi}$ and υ_{zi} in barycentric equatorial system of coordinates are calculated by analogous Equations.

In the calculations, six orbital elements from Table 1, namely, e, a i_e, Ω, ω_e, and M, were used. Other orbital elements were used for testing the calculated data. The perihelion radius R_p and the aphelion radius $R_a = -R_p/(2\alpha_1+1)$ were compared to q and Q, respectively. The orbital period was calculated by Eq. (8) as twice the time of motion from perihelion to aphelion ($r = R_a$). The same Equation was used to calculate the moment at which the asteroid passes the perihelion ($r = r_0$). The calculated values of those quantities were compared to the values of P and t_p given in Table 1. The largest relative difference in terms of q and Q was within $1.9 \cdot 10^{-16}$, and in terms of P and t_p, within $8 \cdot 10^{-9}$.

The coordinates and velocities of the planets and the Moon on epoch JD_0 were calculated by the DE406/LE406 JPL-theory (Ephemerides 2008,

Standish 1998). The masses of those bodies were modified in Grebenikov and Smulsky 2007, and the Apophis mass was calculated assuming the asteroid to be a ball of diameter $d = 270$ m and density $\rho = 3000$ kg/m^3. The masses of all bodies and the initial conditions are given in Table 2.

The starting-data preparation and testing algorithm (3) - (14) was embodied as a MathCad worksheet (program AstCoor2.mcd).

4. Apophis' Encounter with the Planets and the Moon

In the program Galactica, a possibility to determine the minimum distance R_{min} to which the asteroid approaches a celestial body over a given interval ΔT was provided. Here, we integrated Eq. (1) with the initial conditions indicated in Table 2. The integration was performed on the NKS-160 supercomputer at the Computing Center SB RAS, Novosibirsk. In the program Galactica, an extended digit length (34 decimal digits) was used, and for the time step a value $dT = 10^{-5}$ year was adopted. The computations were performed over three time intervals, 0 ÷ 100 years (Figure 1, *a*), 0 ÷ -100 years (Figure 1, *b*), and 0 ÷ 1000 years (Figure 1, *c*).

In the graphs of Figure 1 the points connected with the heavy broken line show the minimal distances R_{min} to which the asteroid approaches the bodies indicated by points embraced by the horizontal line. In other words, a point in the broken line denotes a minimal distance to which, over the time $\Delta T = 1$ year, the asteroid will approach a body denoted by the point in the horizontal line at the same moment. It is seen from Figure 1, *a* that, starting from November 30.0, 2008, over the period of 100 years there will be only one Apophis' approach to the Earth (point A) at the moment T_A = 0.203693547133403 century to a minimum distance R_{minA} = 38907 km. A next approach (point B) will be to the Earth as well, but at the moment T_B = 0.583679164042455 century to a minimum distance R_{minB} = 622231 km, which is 16 times greater than the minimum distance at the first approach. Among all the other bodies, a closest approach with be to the Moon (point D) (see Figure 1, *b*) at T_D = -0.106280550824626 century to a minimum distance R_{minD} =3545163 km.

In the graphs of Figs. 1, *a* and *b* considered above, the closest approaches of the asteroid to the bodies over time intervals $\Delta T = 1$ year are shown. In integrating Eq. (1) over the 1000-year interval (see Figure 1, *c*), we considered the closest approaches of the asteroid to the bodies over time intervals $\Delta T = 10$ years. Over those time intervals, no approaches to Mercury and Mars were identified; in other words, over the 10-year intervals the asteroid closes with other bodies.

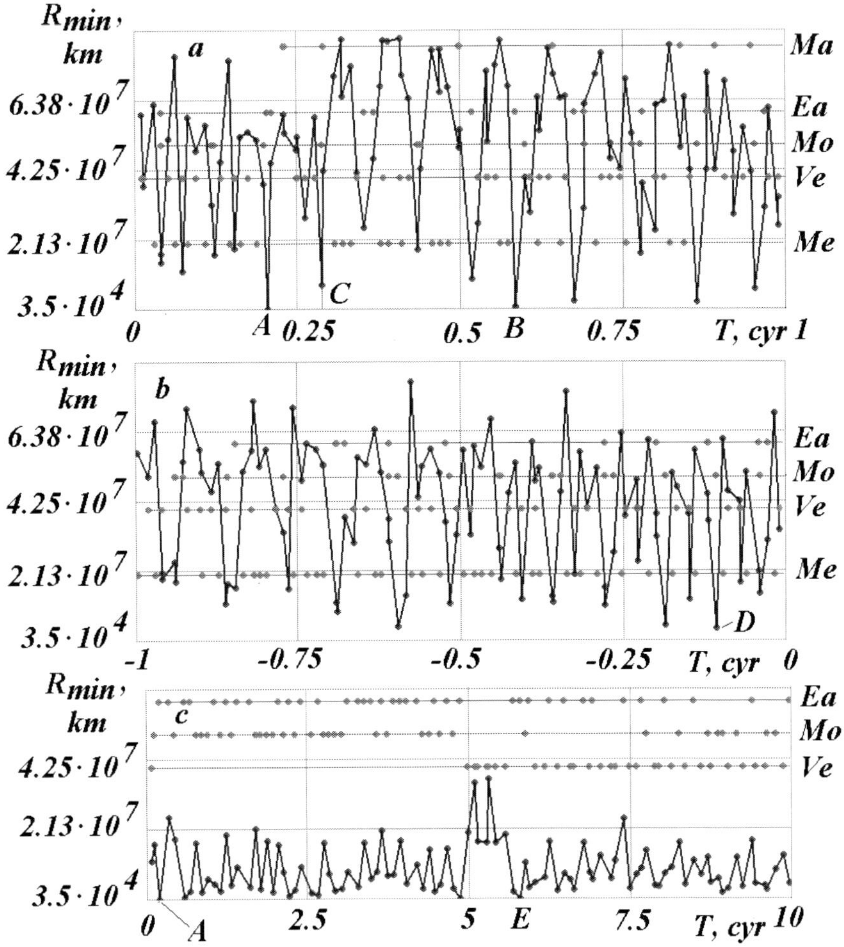

Figure 1. Apophis' encounters with celestial bodies during the time ΔT to a minimum distance R_{min}, km: Mars (Ma), Earth (Ea), Moon (Mo), Venus (Ve) and Mercury (Me); a, b – $\Delta T = 1$ year; c – $\Delta T = 10$ years. T, cyr (1 cyr = 100 yr) is the time in Julian centuries from epoch JD_0 (November 30.0, 2008). Calendar dates of approach in points: A – 13 April 2029; B – 13 April 2067; C – 5 September 2037; E – 10 October 2586.

Like in Figure 1, *a*, there is an approach to the Earth at the moment T_A. A second closest approach is also an approach to the Earth at the point E at T_E = 5.778503 century to a minimum distance R_{minE} = 74002.9 km. During the latter approach, the asteroid will pass the Earth at a minimum distance almost twice that at the moment T_A.

With the aim to check the results, Eq. (1) were integrated over a period of 100 years with double digit length (17 decimal digits) and the same time step, and also with extended digit length and a time step $dT = 10^{-6}$ year. The integration accuracy (see Table 3) is defined (see Melnikov and Smulsky 2009) by the relative change of δM_z, the *z*-projection of the angular momentum of the whole solar system for the 100-year period. As it is seen from Table 3, the quantity δM_z varies from $-4.5 \cdot 10^{-14}$ to $1.47 \cdot 10^{-26}$, i.e., by 12 orders of magnitude. In the last two columns of Table 3, the difference between the moments at which the asteroid most closely approaches the Earth at point A (see Figure 1, *a*) and the difference between the approach distances relative to solution 1 are indicated. In solution 2, obtained with the short digit length, the approach moment has not changed, whereas the minimum distance has reduced by 2.7 m. In solution 3, obtained with ten times reduced integration step, the approach moment has changed by $-2 \cdot 10^{-6}$ year, or by -1.052 minutes. This change being smaller than the step $dT = 1 \cdot 10^{-5}$ for solution 1 and being equal twice the step for solution 3, the value of this change provides a refinement for the approach moment. Here, the refinement for the closest approach distance by -1.487 km is also obtained. On the refined calculations the Apophis approach to the Earth occurs at 21 hours 44 minutes 45 sec on distance of 38905 km. We emphasize here that the graphical data of Figure 1, *a* for solutions *1* and *3* are perfectly coincident. The slight differences of solution *2* from solutions *1* and *3* are observed for $T > 0.87$ century. Since all test calculations were performed considering the parameters of solution *1*, it follows from here that the data that will be presented below are accurate in terms of time within 1', and in terms of distance, within 1.5 km.

At integration on an interval of 1000 years the relative change of the angular momentum is $M_z = 1.45 \cdot 10^{-20}$. How is seen from the solution 1 of Table 3 this value exceeds M_z at integration on an interval of 100 years in 10 times, i.e. the error at extended length of number is proportional to time. It allows to estimate the error of the second approach Apophis with the Earth in T_E = 578 years by results of integrations on an interval of 100 years of the solution with steps $dT = 1 \cdot 10^{-5}$ years and $1 \cdot 10^{-6}$ years. After 88 years from beginning of integration the relative difference of distances between

Apophisом and Earth has become $\delta R_{88} = 1 \cdot 10^{-4}$, that results in an error in distance of 48.7 km in $T_E = 578$ years.

So, during the forthcoming one-thousand-year period the asteroid Apophis will most closely approach the Earth only. This event will occur at the time T_A counted from epoch JD_0. The approach refers to the Julian day JD_A = 2462240.406075 and calendar date April 13, 2029, 21 hour 44'45" GMT. The asteroid will pass at a minimum distance of 38905 km from the Earth center, i.e., at a distance of 6.1 of Earth radii. A next approach of Apophis to the Earth will be on the 578-th year from epoch JD_0; at that time, the asteroid will pass the Earth at an almost twice greater distance.

Table 3. Comparison between the data on Apophis' encounter with the Earth obtained with different integration accuracies: L_{nb} is the digit number in decimal digits

№ solution	L_{nb}	dT, *yr*	δM_z	T_{Ai}-T_{A1}, *yr*	R_{minAi}-R_{minA1}, *km*
1	34	$1 \cdot 10^{-5}$	$1.47 \cdot 10^{-21}$	0	0
2	17	$1 \cdot 10^{-5}$	$-4.5 \cdot 10^{-14}$	0	$-2.7 \cdot 10^{-3}$
3	34	$1 \cdot 10^{-6}$	$1.47 \cdot 10^{-26}$	$-2 \cdot 10^{-6}$	-1.487

The calculated time at which Apophis will close with the Earth, April 13, 2029, coincides with the approach times that were obtained in other reported studies. For instance, in the recent publication Giorgini *et al.* 2008 this moment is given accurate to one minute: 21 hour 45' UTC, and the geocentric distance was reported to be in the range from 5.62 to 6.3 Earth radii, the distance of 6.1 Earth radii falling into the latter range. The good agreement between the data obtained by different methods proves the obtained data to be quite reliable.

As for the possible approach of Apophis to the Earth in 2036, there will be no such an approach (see Figure 1, *a*). A time-closest Apophis' approach at the point *C* to a minimum distance of 7.26 million km will be to the Moon, September 5, 2037.

5. Apophis Orbit Evolution

In integrating motion Eq. (1) over the interval –1 century ≤ T ≤ 1 century the coordinates and velocities of the bodies after a lapse of each one year were recorded in a file, so that a total of 200 files for a one-year time interval were obtained. Then, the data contained in each file were used to integrate Eq. (1) again over a time interval equal to the orbital period of Apophis and, following this, the coordinates and velocities of the asteroid, and those of Sun, were also saved in a new file. These data were used in the program DefTra to determine the parameters of Apophis' orbit relative to the Sun in the equatorial coordinate system. Such calculations were performed hands off for each of the 200 files under the control of the program PaOrb. Afterwards, the angular orbit parameters were recalculated into the ecliptic coordinate system (see Figure 2).

As it is seen from Figure 2, the eccentricity e of the Apophis orbit varies non-uniformly. It shows jumps or breaks. A most pronounced break is observed at the moment T_A, at which Apophis most closely approaches the Earth. A second most pronounced break is observed when Apophis approaches the Earth at the moment T_B.

The longitude of ascending node Ω shows less breaks, exhibiting instead rather monotonic a decrease (see Figure 2). Other orbital elements, namely, i_e, ω_e, a, and P, exhibit pronounced breaks at the moment of Apophis' closest pass near the Earth (at the moment T_A).

The dashed line in Figure 2 indicates the orbit-element values at the initial time, also indicated in Table 1. As it is seen from the graphs, those values coincide with the values obtained by integration of Eq. (1), the relative difference of e, Ω, i_e, ω_e, a, and P from the initial values at the moment $T=0$ (see Table 1) being respectively $9.4\cdot10^{-6}$, $-1.1\cdot10^{-6}$, $3.7\cdot10^{-6}$, $-8.5\cdot10^{-6}$, $1.7\cdot10^{-5}$, and $3.1\cdot10^{-5}$. This coincidence testifies the reliability of computed data at all calculation stages, including the determination of initial conditions, integration of equations, determination of orbital parameters, and transformations between the different coordinate systems.

As it was mentioned in Introduction, apart from non-simplified differential Eq. (1) for the motion of celestial bodies, other equations were also used. It is a well-known fact (see Duboshin 1976) that in perturbed-motion equations orbit-element values are used. For this reason, such equations will yield appreciable errors in determination of orbital-parameter breaks similar to those shown in Figure 2. Also, other solution methods for differential equations exist, including those in which expansions with respect to orbital elements or difference quotients are used. As it was already mentioned in Introduction, these methods proved to be sensitive to various resonance phenomena and sudden orbit changes observed on the approaches between bodies. Eq. (1) and method (2) used in the present study are free of such shortcomings. This suggests that the results reported in the present paper will receive no notable corrections in the future.

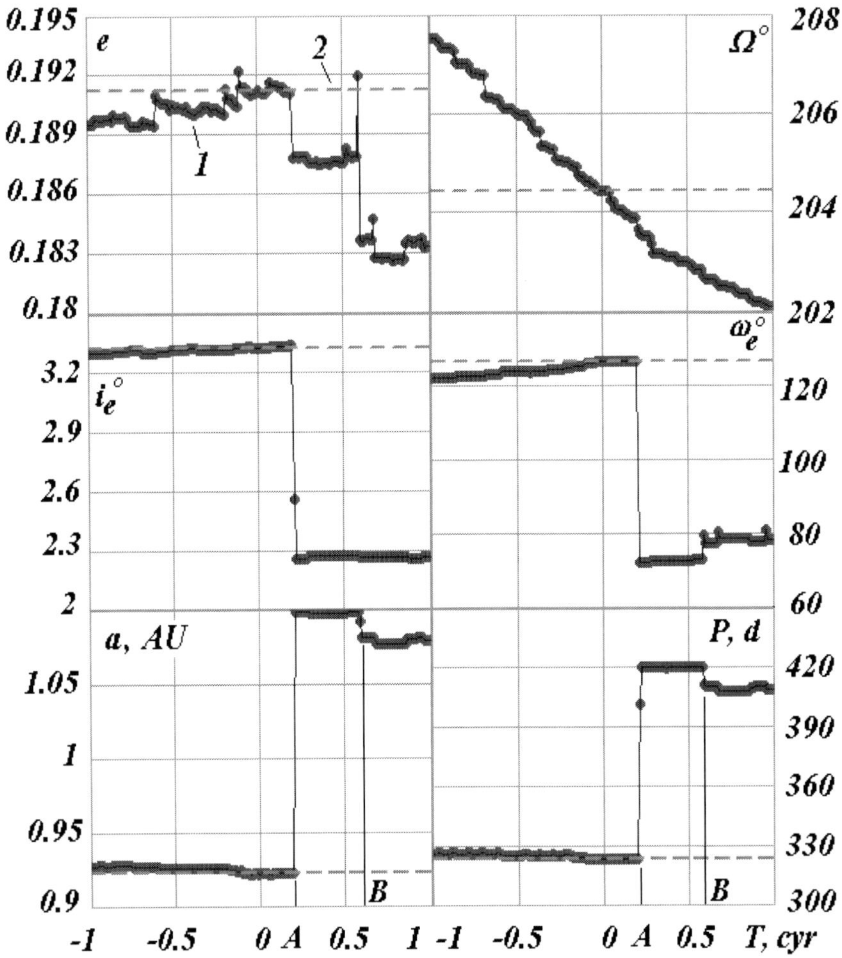

Figure 2. Evolution of Apophis' orbital parameters under the action of the planets, the Moon and the Sun over the time interval -100 years ÷ +100 years from epoch November 30.0, 2008: *1* – as revealed through integration of motion Eq. (1); *2* – initial values according to Table 1. The angular quantities: Ω, i_e, and ω_e are given in degrees; the major semi-axis a in AU; and the orbital period P in days.

6. INFLUENCE OF INITIAL CONDITIONS

With the purpose of check of influence of the initial conditions (IC) on Apophis trajectory the Eq. (1) were else integrated on an interval 100 years with two variants of the initial conditions. The second of variant IC is given on January 04.0, 2010 (see Table 1). They are taken from the JPL Small-Body database 2008 and correspond to the solution with number JPL sol. 144, received Steven R. Chesley on October 23, 2009. In Figure 3 the results of two solutions with various IC are submitted. The line *1* shows the change in time of distance R between Apophis and Earth for 100 years at the first variant IC. As it is seen from the graphs, the distance R changes with oscillations, thus it is possible to determine two periods: the short period $T_{R1} = 0.87$ years and long period T_{R2}. The amplitude of the short period $R_{a1} = 29.3$ million km, and long is $R_{a2} = 117.6$ million km. The value of the long oscillation period up to $T \sim 70$ years is equal $T_{R20} = 7.8$ years, and further it is slightly increased. After approach of April 13, 2029 (point A in Figure 3) the amplitude of the second oscillations is slightly increased. Both short and the long oscillations are not regular; therefore their average characteristics are above given.

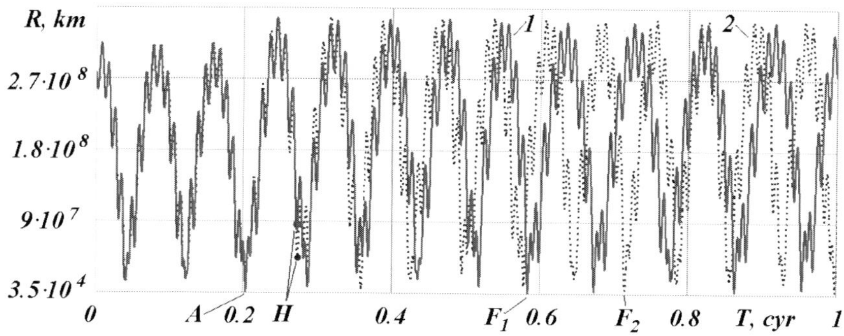

Figure 3. Evolution of distance R between Apophis and Earth for 100 years. Influence of the initial conditions (IC): *1* - IC from November 30.0, 2008; *2* - IC from January 04.0, 2010. Calendar dates of approach in points: A – 13 April 2029; F_1– 13 April 2067; F_2 – 14 April 2080.

Let's note also on the second minimal distance of Apophis approach with the Earth on interval 100 years. It occurs at the time $T_{F1} = 58.37$ years (point F_1 in Figure 3) on distance $R_{F1} = 622$ thousand km. In April 13, 2036 (point H in Figure 3) Apophis passes at the Earth on distance $R_{H1} = 86$ million km. The above-mentioned characteristics of the solution are submitted in Table 4.

The line 2 in Figure 3gives the solution with the second of variant IC with step of integration $dT = 1\cdot10^{-5}$ years. The time of approach has coincided to within 1 minutes, and distance of approach with the second of IC became R_{A2} = 37886 km, i.e. has decreased on 1021 km. To determine more accurate these parameters the Eq. (1) near to point of approach were integrated with a step $dT = 1\cdot10^{-6}$ years. On the refined calculations Apophis approaches with the Earth at 21 hours 44 minutes 53 second on distance R_{A2} = 37880 km. As it is seen from Table 4, this moment of approach differs from the moment of approach at the first of IC on 8 second. As at a step $dT = 1\cdot10^{-6}$ years the accuracy of determination of time is 16 second, it is follows, that the moments of approach coincide within the bounds of accuracy of their calculation.

Table 4. Influence of the initial conditions on results of integration of the Eq. (1) by program Galactica and of the equations of Apophis motion by system Horizons: Time$_A$ and R_{minA} are time and distance of Apophis approach with the Earth in April 13, 2029, accordingly; R_H is distance of passage Apophis with the Earth in April 13, 2036; T_F and R_F are time and distance of the second approach (point *F* on Figure 3)

Parameters	Solutions at different variants of initial conditions					
	Galactica			Horizons		
	1	2	3	1	2	3
	30.11.2008 JPL sol.140	04.01.2010 JPL sol.144	30.11.2008 JPL sol.144	18.07.2006 JPL sol.144	30.11.2008 JPL sol.140	04.01.2010 JPL sol.144
Time$_A$	21:44:45	21:44:53	21:44:45	21:46:47	21:45:47	21:44:45
R_{minA}, km	38905	37880	38813	38068	38161	38068
R_H, 10^6 km	86.0	43.8	81.9	51.9	55.9	51.8
T_F, cyr from 30.11.08	0.5837	0.7138	0.6537	0.4237	0.9437	0.4238
R_F, 10^3 km	622	1663	585	1515	684	1541

The short and long oscillations at two variants IC also have coincided up to the moment of approach. After approach in point *A* the period of long oscillations has decreased up to T_{R22} =7.15 years, i.e. became less than period T_{R20} at the first variant IC. The second approach on an interval 100 years occurs at the moment T_{F2} = 70.28 years on distance R_{F2} =1.663 million km. In 2036 г (point *H*) Apophis passes on distance R_{H2} = 43.8 million km.

At the second variant of the initial conditions on January 04.0, 2010 in comparison with the first of variant the initial conditions of Apophis and of acting bodies are changed. To reveal only errors influence of Apophis IC, the third variant of IC is given (see Table 1) as first of IC on November 30.0, 2008, but the Apophis IC are calculated in system Horizons according to JPL sol. 144. How follows from Table 1, from six elements of an orbit e, a, i_e, Ω, ω_e and M the differences of three ones: i_e, Ω и ω_e from similar elements of the first variant of IC are 2.9, 1.6 and 1.5 appropriate uncertainties. The difference of other elements does not exceed their uncertainties.

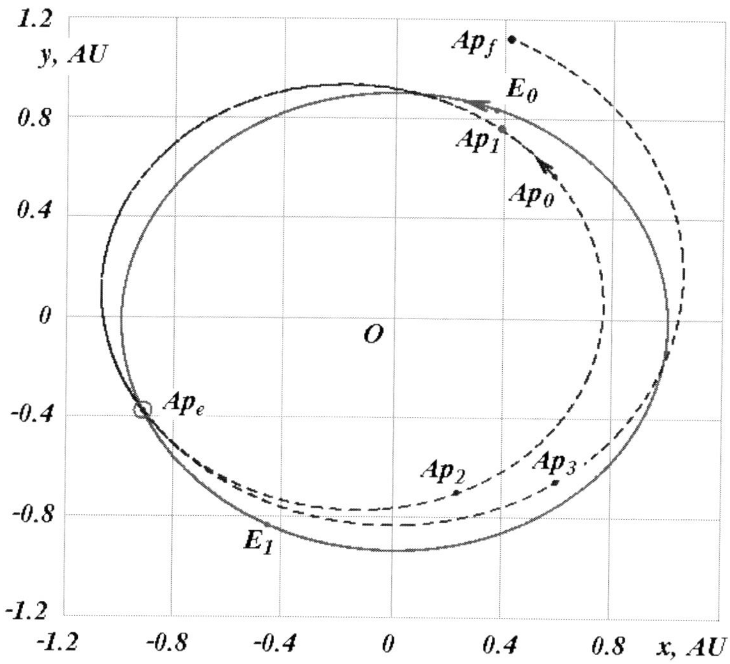

Figure 4. The trajectories of Apophis (Ap) and Earth (E) in the barycentric equatorial coordinate system xOy over a two-year period: Ap_0 and E_0 are the initial position of Apophis and Earth; Ap_f is the end point of the Apophis trajectory; Ap_e is the point at which Apophis most closely approaches the Earth; the coordinates x and y are given in AU.

At the third variant of IC with step of integration $dT = 1{\cdot}10^{-5}$ year the moment of approach has coincided with that at the first variant of IC. The distance of approach became $R_{A3} = 38814$ km, i.e. has decreased on 93 km. For more accurate determination of these parameters the Eq. (1) near to a point of approach were also integrated with a step $dT = 1{\cdot}10^{-6}$ year. On the refined calculations at the third variant of IC Apophis approaches with the Earth at 21 hours 44 minutes 45 second on distance $R_{A3} = 38813$ km. These and other characteristics of the solution are given in Table 4. In comparison with the first variant IC it is seen, that distance of approach in 2036 and parameters of the second approach in point F_1 are slightly changed. The evolution of distance R in a Figure 3 up to $T = 0.6$ centuries practically coincides with the first variant (line *1*).

It is seen (Table 4) that the results of the third variant differ from the first one much less than from the second variant. In the second variant the change of positions and velocities of acting bodies since November 30, 2008 for 04.01.2010 is computed under DE406, and in the third variant it does under the program Galactica. The initial conditions for Apophis in two variants are determined according to alike JPL sol. 144, i.e. in these solutions the IC differ for acting bodies. As it is seen from Table 4, the moment of approach in solutions *2* and *3* differs on 8 seconds, and the approach distance differs on 933 km. Other results of the third solution also differ in the greater degree with second ones, in comparison of the third solution with first one. It testifies that the differences IC for Apophis are less essential in comparison with differences of results of calculations under two programs: Galactica and DE406 (or Horizons).

So, the above-mentioned difference of the initial conditions (variants *1* and *3* tab. *4*) do not change the time of approach of April 13, 2029, and the distance of approach in these solutions differ on 102 km. Other characteristics: R_H, T_F and R_F also change a little. Therefore it is possible to make a conclusion, that the further refinement of Apophis IC will not essentially change its trajectory.

The same researches on influence of the initial conditions we have carried out with the integrator of NASA. In system Horizons (the JPL Horizons On-Line Ephemeris System, manual look on a site http://ssd.jpl.nasa.gov/?horizons_doc) there is opportunity to calculate asteroid motion on the same standard dynamic model (SDM), on which the calculations in Giorgini *et al.* 2008 are executed. Except considered two IC we used one more IC for Apophis at date of July 12, 2006, which is close to date of September 01, 2006 in Giorgini *et al.* 2008. The characteristics and basic results of all solutions are given in Table 4. In these solutions the similar results are received. For example, for *3*-rd variant of Horizons the graphs R in a Figure 3 up to $T = 0.45$ centuries practically has coincided with *2*-nd variant of Galactica. The time of approach in April 13, 2029 changes within the bounds of 2 minutes, and the distance is close to 38000 km. The distance of approach in April 13, 2036 changes from 52 up to 56 million km. The characteristics of second approach for 100 years changes in the same bounds, as for the solutions on the program

Galactica. The above-mentioned other relations about IC influence have also repeated for the NASA integrator.

So, the calculations at the different initial conditions have shown that Apophis in 2029 will be approached with the Earth on distance 38÷39 thousand km, and in nearest 100 years it once again will approach with the Earth on distance not closer 600 thousand km.

7. Examination of Apophis' Trajectory in the Vicinity of Earth

In order to examine the Apophis trajectory in the vicinity of Earth, we integrated Eq. (1) over a two-year period starting from T_1 = 0.19 century. Following each 50 integration steps, the coordinate and velocity values of Apophis and Earth were recorded in a file. The moment T_A at which Apophis will most closely approach the Earth falls into this two-year period. The ellipse E_0E_1 in Figure 4 shows the projection of the two-year Earth's trajectory onto the equatorial plane xOy. Along this trajectory, starting from the point E_0, the Earth will make two turns. The two-year Apophis trajectory in the same coordinates is indicated by points denoted with the letters Ap. Starting from the point Ap_0, Apophis will travel the way $Ap_0Ap_1Ap_eAp_2Ap_0Ap_1$ to most closely approach the Earth at the point Ap_e at the time T_A. After that, the asteroid will follow another path, namely, the path $Ap_eAp_3Ap_f$.

Figure 5, *a* shows the trajectory of Apophis relative to the Earth. Here, the relative coordinates are determined as the difference between the Apophis (Ap) and Earth (E) coordinates:

$$y_r = y_{Ap} - y_E;\ x_r = x_{Ap} - x_E. \qquad (15)$$

Along trajectory 1, starting from the point Ap_0, Apophis will travel to the Earth-closest point Ap_e, the trajectory ending at the point Ap_f. The loops in the Apophis trajectory represent a reverse motion of Apophis with respect to Earth. Such loops are made by all planets when observed from the Earth (Smulsky 2007).

At the Earth-closest point Ap_e the Apophis trajectory shows a break. In Figure 5, *b* this break is shown on a larger scale. Here, the Earth is located at the origin, point 2. The Sun (see Figure 4) is located in the vicinity of the barycenter O, i.e., in the upper right quadrant of the Earth-closest point Ap_e.

Hence, the Earth-closest point will be passed by Apophis as the latter will move in between the Earth and the Sun (see Figure 5, *b*). As it will be shown below, this circumstance will present certain difficulties for possible use of the asteroid.

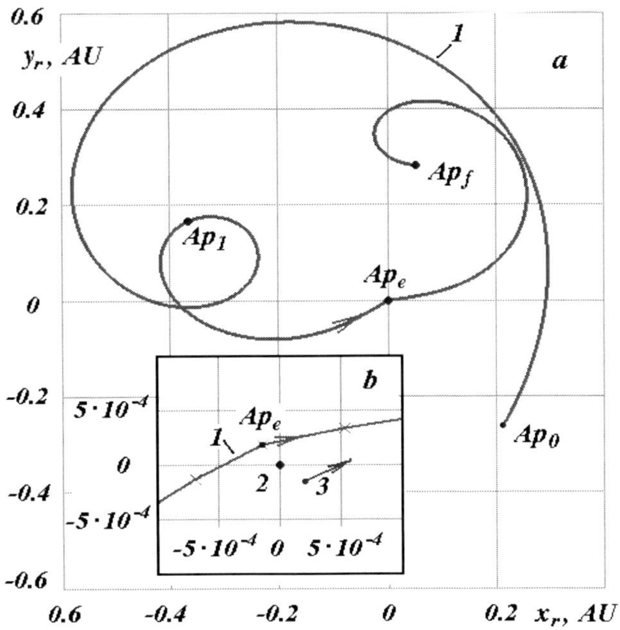

Figure 5. Apophis' trajectory (1) in the geocentric equatorial coordinate system x_rOy_r: *a* – on the normal scale, *b* – on magnified scale on the moment of Apophis' closest approaches to the Earth (2); *3* – Apophis' position at the moment of its closest approach to the Earth following the correction of its trajectory with factor $k = 0.9992$ at the point Ap_1; the coordinates x_r and y_r are given in AU.

8. Possible Use of Asteroid Apophis

So, on April 13, 2029, we will become witnesses of a unique phenomenon, the pass of a body 31 million tons in mass near the Earth at a minimum distance of 6 Earth radii from the center of Earth. Over subsequent 1000 years, Apophis will never approach our planet closer.

Many pioneers of cosmonautics, for instance, K.E. Tsiolkovsky, Yu.A. Kondratyuk, D.V. Cole etc. believed that the near-Earth space will be explored using large manned orbital stations. Yet, delivering heavy masses from Earth into orbit presents a difficult engineering and ecological problem. For this reason, the lucky chance to turn the asteroid Apophis into an Earth bound satellite and, then, into a habited station presents obvious interest.

Among the possible applications of a satellite, the following two will be discussed here. First, a satellite can be used to create a space lift. It is known that a space lift consists of a cable tied with one of its ends to a point at the Earth equator and, with the other end, to a massive body turning round the Earth in the equatorial plane in a 24-hour period, $P_d = 24{\cdot}3600$ sec. The radius of the satellite geostationary orbit is

$$R_{gs} = \sqrt[3]{P_d^2 G(m_A + m_E)/4\pi^2} = 42241 \text{ km} = 6.62\ R_{Ee} \qquad (16)$$

In order to provide for a sufficient cable tension, the massive body needs to be spaced from the Earth center a distance greater than R_{gs}. The cable, or several such cables, can be used to convey various goods into space while other goods can be transported back to the Earth out of space.

If the mankind will become able to make Apophis an Earth bound satellite and, then, deflect the Apophis orbit into the equatorial plane, then the new satellite would suit the purpose of creating a space lift.

A second application of an asteroid implies its use as a “shuttle” for transporting goods to the Moon. Here, the asteroid is to have an elongated orbit with a perihelion radius close to that of a geostationary orbit and an apogee radius approaching the perigee radius of the lunar orbit. In the latter case, at the geostationary-orbit perigee goods would be transferred onto the satellite Apophis and then, at the apogee, those goods would arrive at the Moon.

The two applications will entail the necessity of solving many difficult problems which now can seem even unsolvable. On the other hand, none of those problems will be solved at all without making asteroid an Earth satellite. Consider now the possibilities available here.

The velocity of the asteroid relative to the Earth at the Earth-closest point Ap_e is v_{AE} =7.39 km s^{-1}. The velocity of an Earth bound satellite orbiting at a fixed distance R_{minA} from the Earth (circular orbit) is

$$v_{CE} = \sqrt{G(m_A + m_E)/R_{\min A}} = 3.2 \text{ km s}^{-1} \qquad (17)$$

For the asteroid to be made an Earth-bound satellite, its velocity v_{AE} should be brought close to v_{CE}. We performed integration of Eq. (1) assuming the Apophis velocity at the moment T_A to be reduced by a factor of 1.9, i.e., the velocity v_{AE} =7.39 km s^{-1} at the moment T_A was decreased to 3.89 km s^{-1}. In the later case, Apophis becomes an Earth bound satellite with the following orbit characteristics: eccentricity e_{s1} = 0.476, equator-plane inclination angle i_{s1} = 39.2°, major semi-axis a_{s1} = 74540 km, and sidereal orbital period P_{s1} = 2.344 days.

We examined the path evolution of the satellite for a period of 100 years. In spite of more pronounced oscillations of the orbital elements of the satellite in comparison with those of planetary orbit elements, the satellite's major semi-axis and orbital period proved to fall close to the indicated values. For the relative variations of the two quantities, the following estimates were obtained: $|\delta a| < \pm 2.75 \cdot 10^{-4}$ and $|\delta P| < \pm 4.46 \cdot 10^{-4}$. Yet, the satellite orbits in a direction opposite both to the Earth rotation direction and the direction of Moon's orbital motion. That is why the two discussed applications of such a satellite turn to be impossible.

Thus, the satellite has to orbit in the same direction in which the Earth rotates. Provided that Apophis (see Figure 5, *b*) will round the Earth from the night-side (see point *3*) and not from the day-side (see line *1*), then, on a decrease of its velocity the satellite will be made a satellite orbiting in the required direction.

For this matter to be clarified, we have integrated Eq. (1) assuming different values of the asteroid velocity at the point Ap_1 (see Figure 5). This point, located at half the turn from the Earth-closest point Ap_e, will be passed by Apophis at the time T_{Ap1}=0.149263369488169 century. At the point Ap_1 the projections of the Apophis velocity in the barycentric equatorial coordinate system are v_{Ap1x} = -25.6136689 km s^{-1}, v_{Ap1y} = 17.75185451 km s^{-1}, and v_{Ap1z} = 5.95159206 km s^{-1}. In the numerical experiments, the component values of the satellite velocity were varied to one and the same proportion by multiplying all them by a single factor *k*, and then Eq. (1) were integrated to determine the trajectory of the asteroid. Figure 6 shows the minimum distance to which Apophis will approach the Earth versus the value of *k* by which the satellite velocity at the point Ap_1 was reduced.

We found that, on decreasing the value of *k* (see Figure 6), the asteroid will more closely approach the Earth, and at *k* = 0.9999564 Apophis will collide with the Earth. On further decrease of asteroid velocity the asteroid will close with the Earth on the Sun-opposite side, and at k = 0.9992 the asteroid will approach the Earth center (point 3 in Figure 5, *b*) to a minimum distance R_{min3} = 39157 km at the time T_3 = 0.2036882 century. This distance R_{min3} roughly equals the distance R_{minA} to which the asteroid was found to approach the Earth center while moving in between the Earth and the Sun.

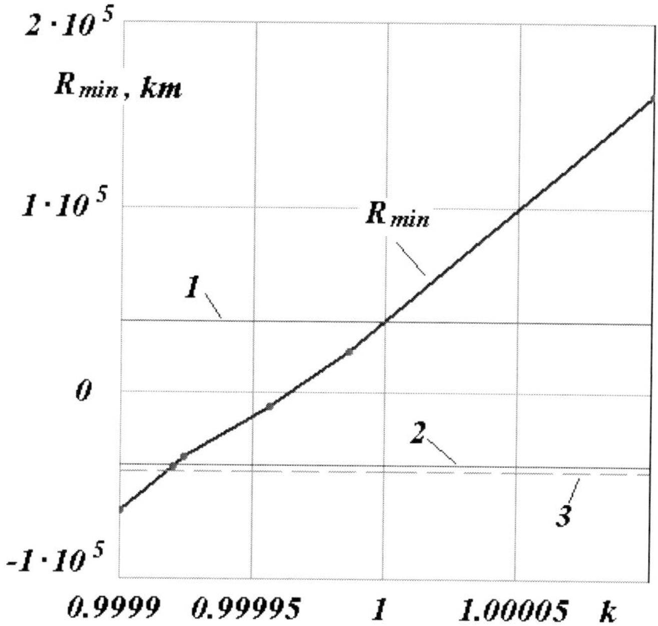

Figure 6. The minimum distance R_{min} to which Apophis will approach the Earth center versus the value of k (k is the velocity reduction factor at the point A_{p1} (see Figure 4)). The positive values of R_{min} refer to the day-side: the values of R_{min} are given in km; *1* – the minimum distance to which Apophis will approach the Earth center on April 13, 2029 (day-side); *2* – the minimum distance to which Apophis will approach the Earth center after the orbit correction (night-side); *3* – geostationary orbit radius R_{gs}.

In this case, the asteroid velocity relative to the Earth is also v_{AE} =7.39 km s^{-1}. On further decrease of this velocity by a factor of 1.9, i.e., down to 3.89 km s^{-1} Apophis will become an Earth bound satellite with the following orbit parameters: eccentricity e_{s2} = 0.486, equator plane inclination angle i_{s2} = 36°, major semi-axis a_{s2} = 76480 km, and sidereal period P_{s2} = 2.436 day. In addition, we investigated into the path evolution of the Earth bound satellite over a 100-year period. The orbit of the satellite proved to be stable, the satellite orbiting in the same direction as the Moon does.

Thus, for Apophis to be made a near-Earth satellite orbiting in the required direction, two decelerations of its velocity need to be implemented. The first deceleration is to be effected prior to the Apophis approach to the Earth, for instance, at the point Ap_1 (see Figure 4), 0.443 year before the Apophis approach to the Earth. Here, the Apophis velocity needs to be decreased by 2.54 m/s. A second deceleration is to be effected at the moment the asteroid closes with the Earth. In the case under consideration, in which the asteroid moves in an elliptic orbit, the asteroid velocity needs to be decreased by 3.5 km s^{-1}.

Slowing down a body weighing 30 million tons by 3.5 km s^{-1} is presently a difficult scientific and engineering problem. For instance, in Rykhlova *et al.* 2007 imparting Apophis with a velocity of 10^{-6} m/s was believed to be a problem solvable with presently available engineering means. On the other hand, Rykhlova *et al.* 2007 consider increasing the velocity of such a body by about 1-2 cm/s a difficult problem. Yet, with Apophis being on its way to the Earth, we still have a twenty-year leeway. After the World War II, even more difficult a problem, that on injection of the first artificial satellite in near-Earth orbit and, later, the launch of manned space vehicles, was successfully solved in a period of ten years. That is why we believe that, with consolidated efforts of mankind, the objective under discussion will definitely be achieved.

It should be emphasized that the authors of Giorgini *et al.* 2008 considered the possibility of modifying the Apophis orbit for organizing its impact onto asteroid (144898) 2004 VD17. There exists a small probability of the asteroid's impact onto the Earth in 2102. Yet, the problem on reaching a required degree of coordination between the motions of the two satellites presently seems to be hardly solvable. This and some other examples show that many workers share an opinion that substantial actions on the asteroid are necessary for making the solution of the various space tasks a realistic program.

9. Asteroid 1950 DA Approaches to the Earth

The distances to which the asteroid 1950 DA will approach solar-system bodies are shown versus time in Figure 7. It is seen from Figure 7, *a*, that, following November 30.0, 2008, during the subsequent 100-year period the asteroid will most closely approach the Moon: at the point *A* (T_A=0.232532 cyr and R_{min}=11.09 million km) and at the point *B* (T_B=0.962689 cyr and R_{min}= 5.42 million km). The encounters with solar-system bodies the asteroid had over the period of 100 past years are shown in Figure 7, *b*. The asteroid most closely approached the Earth twice: at the point *C* (T_C = -0.077395 cyr and R_{min}=7.79 million km), and at the point *D* (T_D=-0.58716 cyr and R_{min}=8.87 million km).

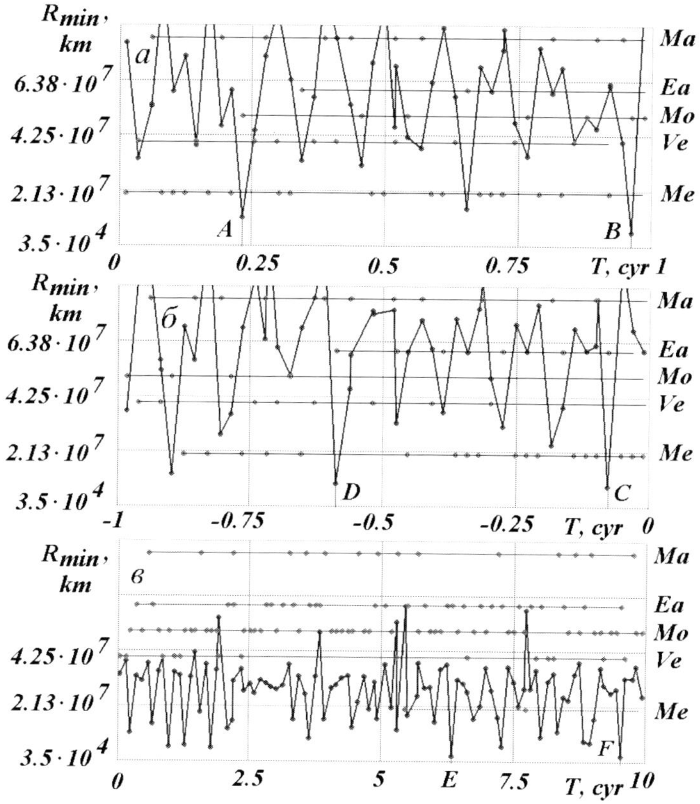

Figure 7. Approach of the asteroid 1950 DA to solar-system bodies. The approach distances are calculated with time interval ΔT: *a*, *b* – ΔT=1 year; *c* – ΔT = 10 years. R_{min}, km is the closest approach distance. Calendar dates of approach in points see Table 5. For other designations, see Figure 1.

Over the interval of forthcoming 1000 years, the minimal distances to which the asteroid will approach solar-system bodies on time span ΔT=10 years are indicated in Figure 7, *c*. The closest approach of 1950 DA will be to the Earth: at the point *E* (T_E = 6.322500 cyr and R_{min}=2.254 million km), and at the point *F* (T_F = 9.532484 cyr and R_{min}=2.248 million km).

To summarize, over the 1000-year time interval the asteroid 1950 DA will most closely approach the Earth twice, at the times T_E and T_F, to a minimum distance of 2.25 million km in both cases. The time T_E refers to the date March 6, 2641, and the time T_F, to the date March 7, 2962.

Table 5. Comparison between the data on asteroid 1950 DA encounters with the Earth and Moon: our data are denoted with characters A, B, C, D, E, F, as in Figure 7, and the data by Giorgini et al. [24] are denoted as Giorg

Source	JD, days	Date	Time, days	Body	R_{min}, AU
D	2433354	1950-03-13	0.730	Earth	0.059273
Giorg.	-	1950-03-12	0.983	Earth	0.059286
C	2451973	2001-03-05	0.157	Earth	0.052075
Giorg.	-	2001-03-05	0.058	Earth	0.052073
A	2463293	2032-03-02	0.222	Moon	0.074158
Giorg.	-	2032-03-02	0.281	Earth	0.075751
B	2489962	2105-03-09	0.224	Moon	0.036260
Giorg.	-	2105-03-10	0.069	Earth	0.036316
E	2685729	2641-03-06	0.338	Earth	0.015070
Giorg.	-	2641-03-14	0.330	Earth	0.015634
F	2802974	2962-03-07	0.985	Earth	0.015030
Giorg.	-	2880-03-16	0.836	Earth	0.001954

Giorgini et al. 2002 calculated the nominal 1950 DA trajectory using earlier estimates for the orbit-element values of the asteroid, namely, the values by the epoch of March 10.0, 2001 (JPL sol. 37). In Giorgini et al. 2002, as the variation of initial conditions for the asteroid, ranges were set three times wider than the uncertainty in element values. For the extreme points of the adopted ranges, in the calculations 33 collision events were registered. In this connection, Giorgini et al. 2002 have entitled their publication «Asteroid 1950 DA Encounter with Earth in 2880…».

We made our calculations using the orbit-element values of 1950 DA by the epoch of November 30.0, 2008 (JPL sol. 51) (see Table 1). By system Horizons the JPL sol. 37 can be prolonged till November 30.0, 2008. As it is seen in this case, the difference of orbital elements of the solution 37 from the solution 51 on two - three order is less, than uncertainties of orbit elements, i.e. the orbital elements practically coincide.

With the aim to trace how the difference methods of calculation has affected the 1950 DA motion, in Table 5 we give a comparison of the approach times of Figure 7 with the time-closest approaches predicted in Giorgini et al. 2002. According to Table 5, the shorter the separation between the approach times (see points *C* and *A*) and the start time of calculation (2008-11-30), the better is the coincidence in terms of approach dates and minimal approach distances R_{min}. For more remote times (see points *D* and *B*) the approach times differ already by 1 day. At the point *E*, remote from the start time of calculation by 680 year, the approach times differ already by eight days, the approach distances still differing little. At the most remote point *F*, according to our calculations, the asteroid will approach the Earth in 2962 to a distance of 0.015 AU, whereas, according to the data of Giorgini et al. 2002, a most close approach to the Earth, to a shorter distance, will be in 2880.

So, our calculations show that the asteroid 1950 DA will not closely approach the Earth. It should be noted that our calculation algorithm for predicting the motion of the asteroid differs substantially from that of Giorgini et al. 2002. We solve non-simplified Eq. (1) by a high-precision numerical method. In doing so, we take into account the Newtonian gravitational interaction only. In Giorgini et al. 2002, additional weak actions on the asteroid were taken into account. Yet, the position of celestial bodies acting on the asteroid is calculated from the ephemerides of DE-series. Those ephemeredes approximate observational data and, hence, they describe those data to good precision. Yet, the extent to which the predicted motion of celestial bodies deviates from the actual motion of these bodies is the greater the farther the moment of interest is remote from the time interval during which the observations were made. We therefore believe that the difference between the present calculation data for the times 600 and 900 years (points *E* and *F* in Table 5) and the data of Giorgini et al. 2002 results from the indicated circumstance.

10. Evolution of the 1950 DA Orbit

Figure 8 shows the evolution of 1950 DA orbital elements over a 1000-year time interval as revealed in calculations made with time span ΔT=10 years. With the passage of time, the orbit eccentricity e non-monotonically increases. The angle of longitude of ascending node Ω, the angle of inclination i_e to the ecliptic plane, and the angle of perihelion argument ω_e show more monotonic variations. The semi-axis a and the orbital period P both oscillate about some mean values. As it is seen from Figure 8, at the moments of encounter with the Earth, T_E and T_F, the semi-axis a and the period P show jumps. At the same moments, all the other orbit elements exhibit less pronounced jumps.

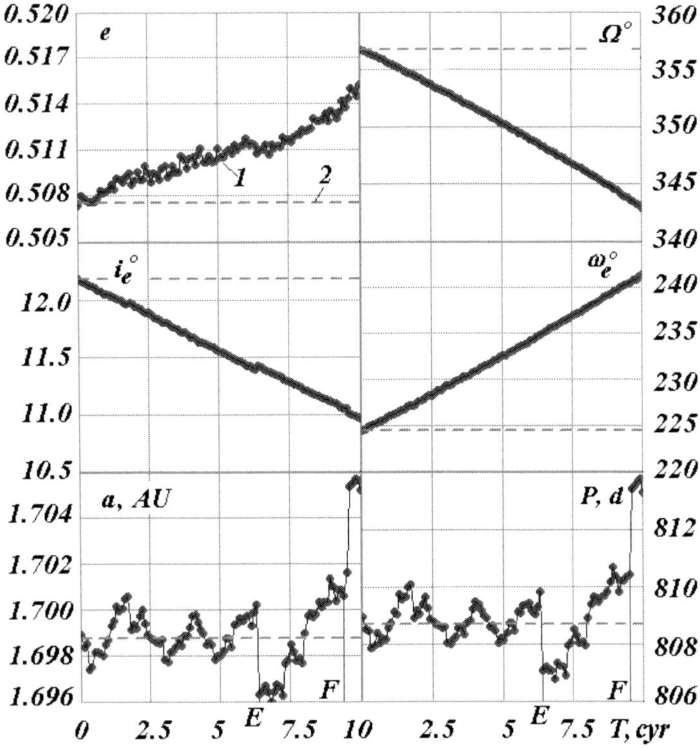

Figure 8. Evolution of 1950 DA orbital parameters under the action of the planets, the Moon, and the Sun over the time interval 0÷1000 from the epoch November 30.0, 2008: *1*- as revealed through integration of motion equation (1) obtained with the time interval $\Delta T = 10$ years: *2* – initial values according to Table 1. The angular quantities, Ω, i_e, and ω_e, are given in degrees, the major semi-axis a – in *AU,* and the orbital period P, in days.

The dashed line in Figure 8 indicates the initial-time values of orbital elements presented in Table 1. As it is seen from the graphs, these values are perfectly coincident with the values for T=0 obtained by integration of Eq. (1). The relative differences between the values of e, Ω, i_e, ω_e, a, and P and the initial values of these parameters given in Table 1 are $-3.1\cdot10^{-4}$, $-1.6\cdot10^{-5}$, $-6.2\cdot10^{-5}$, $-1.5\cdot10^{-5}$, $-1.5\cdot10^{-5}$, $-1.0\cdot10^{-4}$, and $-3.0\cdot10^{-4}$, respectively. Such a coincidence validates the calculations at all stages, including the determination of initial conditions, integration of Eq. (1), determination of orbital-parameter values, and the transformation between different coordinate systems.

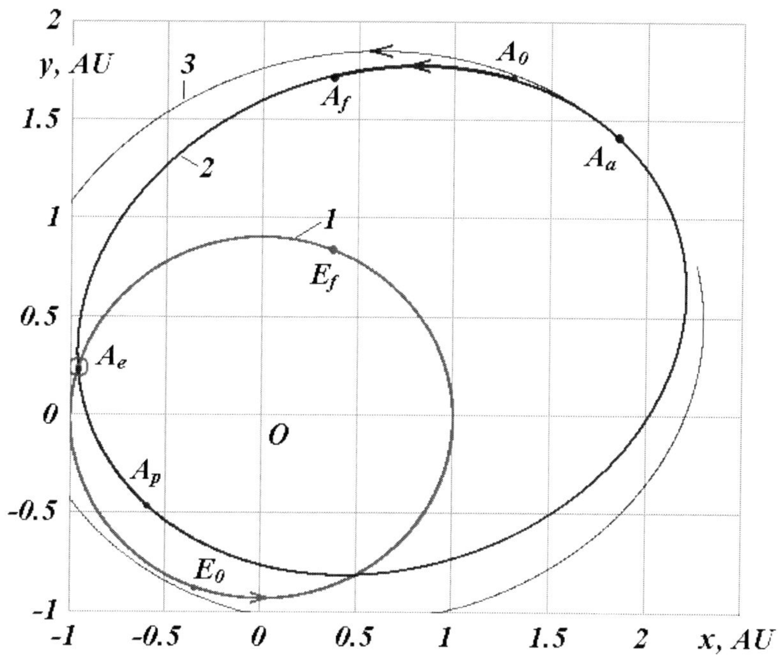

Figure 9. The trajectories of Earth (*1*) and 1950 DA (*2*) in the barycentric equatorial coordinate system *xOy* over 2.5 years in the encounter epoch of March 6, 2641 (point A_e): A_0 and E_0 are the starting points of the 1950 DA and Earth trajectories; A_f and E_f are the end points of the 1950 DA and Earth trajectories; *3* - 1950 DA trajectory after the correction applied at the point A_a is shown arbitrarily; the coordinates *x* and *y* are given in AU.

It should be noted that the relative difference for the same elements of Apophis is one order of magnitude smaller. The cause for the latter can be explained as follows. Using the data obtained by integrating Eq. (1), we determine the orbit elements at the time equal to half the orbital period. Hence, our elements are remote from the time of determination of the initial conditions by that time interval. Since the orbital period of Apophis is shorter than that of 1950 DA, the time of determination of Apophis' elements is 0.66 year closer in time to the time of determination of initial conditions than the same time for 1950 DA.

10. STUDY OF THE 1950 DA TRAJECTORY IN THE ENCOUNTER EPOCH OF MARCH 6, 2641

Since the distances to which the asteroid will approach the Earth at the times T_E and T_F differ little, consider the trajectories of the asteroid and the Earth at the nearest approach time T_E, March 6, 2641. The ellipse E_0E_f in Figure 9 shows the projection of the Earth trajectory over a 2.5-year period onto the equatorial plane *xOy*. This projection shows that, moving from the point E_0 the Earth will make 2.5 orbital turns. The trajectory of 1950 DA starts at the point A_0. At the point A_e the asteroid will approach the Earth in 2641 to a distance of 0.01507 AU. The post-encounter trajectory of the asteroid remains roughly unchanged. Then, the asteroid will pass through the perihelion point A_p and aphelion point A_a, and the trajectory finally ends at the point A_f.

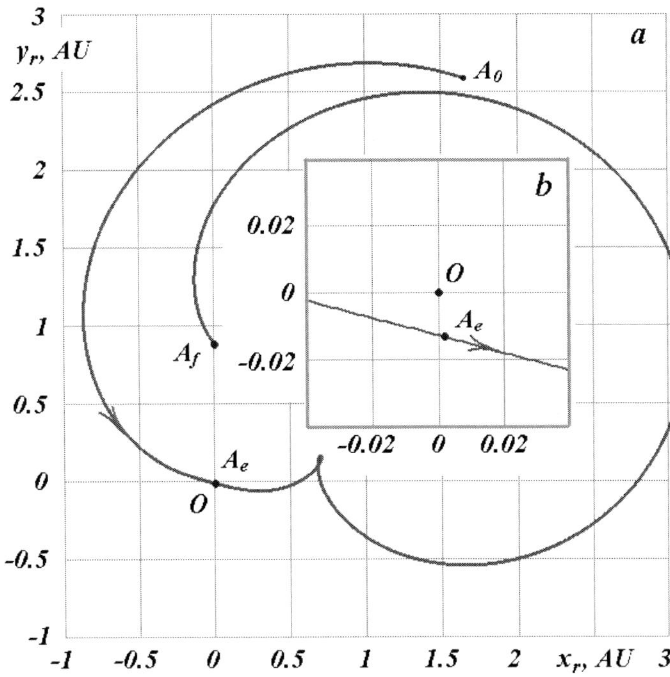

Figure 10. The 1950 DA trajectory in the geocentric equatorial coordinate system x_rOy_r: *a* – on ordinary scale; *b* – on an enlarged scale by the moment of 1950 DA encounter with the Earth: point *O* – the Earth, point A_e – the asteroid at the moment of its closest approach to the Earth; the coordinates x_r and y_r are given in AU.

Figure 10, *a* shows the trajectory of the asteroid relative to the Earth. The relative coordinates x_r and y_r were calculated by a Equation analogous to (15). Starting at the point A_0, the asteroid 1950 DA will move to the point A_e, where it will most closely approach the Earth, the end point of the trajectory being the point A_f. The loop in the 1950 DA trajectory represents a reverse motion of the asteroid relative to the Earth.

On an enlarged scale, the encounter of the asteroid with the Earth is illustrated by Figure 10, *b*. The Sun is in the right upper quadrant. The velocity of the asteroid relative to the Earth at the closing point A_e is v_{AE}=14.3 km s^{-1}.

12. MAKING THE ASTEROID 1950 DA AN EARTH-BOUND SATELLITE

Following a deceleration at the point A_e (see Figure 10, *b*), the asteroid 1950 DA can become a satellite orbiting around the Earth in the same direction as the Moon does. At this point E (see Table 5) the distance from the asteroid to the Earth's center is R_{minE} = 2.25 million km, the mass of the asteroid being m_A = 1.57 milliard ton. According to (17), the velocity of a satellite moving in a circular orbit of radius R_{minE} is v_{CE}=0.421 km s^{-1}. For the asteroid 1950 DA to be made a satellite, its velocity needs to be brought close to the value v_{CE} or, in other words, the velocity of the asteroid has to be decreased by $\Delta V\approx$13.9 km s^{-1}. In this situation, the asteroid's momentum will become decreased by a value $m_a\Delta V$=2.18·10^{16} kg·m/s, for Apophis the same decrease amounts to $m_a\cdot\Delta V$=1.08 10^{14} kg·m s^{-1}, a 200 times greater value. Very probably, satellites with an orbital radius of 2.25 million km will not find a wide use. In this connection, consider another strategy for making the asteroid an Earth-bound satellite. Suppose that the velocity of the asteroid at the aphelion of its orbit (point A_a in Figure 9) was increased so that the asteroid at the orbit perihelion has rounded the Earth orbit on the outside of it passing by the orbit at a distance R_1. To simplify calculations, we assume the Earth's orbit to be a circular one with a radius equals the semi-axis of the Earth orbit a_E = 1 AU. So, in the corrected orbit of the asteroid the perihelion radius will be

$$R_{pc} = a_E + R_1. \qquad (18)$$

Then, let us decrease the velocity of the asteroid at the perihelion of the corrected orbit to a value such that to make the asteroid an Earth-bound satellite. To check efficiency of this strategy, perform required calculations based on the two-body interaction model for the asteroid and the Sun (Smulsky 2007, Smulsky 2008). We write the expression for the parameter of trajectory in three forms:

$$\alpha_1 = -0.5(1 + R_p / R_a) = \frac{\mu_1}{R_p \cdot v_p^2} = \frac{R_p \mu_1}{R_a^2 \cdot v_a^2}, \tag{19}$$

where

$$\mu_I = -G\,(m_s + m_{As}) \tag{20}$$

is the interaction parameter of the Sun and the asteroid, m_S is the Sun mass, m_{As} is the asteroid mass, and α_I= -0.6625 is the 1950 DA trajectory parameter.

Then, using (19), for the corrected orbit of the asteroid with parameters R_{pc} and v_{ac} we obtain:

$$-0.5(1 + R_{pc} / R_a) = \frac{R_{pc} \mu_1}{R_a^2 v_{ac}^2}. \tag{21}$$

From (21), we obtain the corrected velocity of the asteroid at aphelion:

$$v_{ac} = \sqrt{\frac{2 \cdot R_{pc}(-\mu_1)}{R_a^2 (R_a + R_{pc})}}. \tag{22}$$

Using (19), we express μ_I in terms of α_I and v_a, and after substitution of this expression into (22) we obtain the corrected velocity at aphelion:

$$v_{ac} = v_a \sqrt{\frac{2(-\alpha_1) R_{pc} \cdot R_a}{(R_a + R_{pc}) \cdot R_p}}. \tag{23}$$

From the second Kepler law, $R_a \cdot v_{ac} = R_{pc} \cdot v_{pc}$, we determine the velocity at the perihelion of the corrected orbit:

$$v_{pc} = v_{ac} \cdot R_a / R_{pc} \,. \qquad (24)$$

As a numerical example, consider the problem on making the asteroid 1950 DA an Earth-bound satellite with a perihelion radius equal to the geostationary orbit radius $R_1 = R_{gs} = 42241$ km. Prior to the correction, the aphelion velocity of the asteroid is $v_a = 13.001$ km s^{-1}, whereas the post-correction velocity calculated by Equation (23) is $v_{ac} = 13.912$ km s^{-1}. Thus, for making the asteroid a body rounding the Earth orbit it is required to increase its velocity at the point A_a in Figure 9 by 0.911 km s^{-1}. The corrected orbit is shown in Figure 9 with line *3*.

According to (24), the velocity of the asteroid at the perihelion of the corrected orbit is $v_{pc} = 35.622$ km s^{-1}. Using Eq. (7), for a circular Earth orbit with $\alpha_l = -1$ and $R_p = a_E$, and with the asteroid mass m_{AS} replaced with the Earth mass m_E, for the orbital velocity of the Earth we obtain a value $v_{OE} = 29.785$ km s^{-1}. According to (17), the velocity of the satellite in the geostationary orbit is $v_{gs} = 3.072$ km s^{-1}. Since those velocities add up, for the asteroid to be made an Earth satellite, its velocity has to be decreased to the value $v_{OE} + v_{CE} = 32.857$ km s^{-1}. Thus, the asteroid 1950 DA will become a geostationary satellite following a decrease of its velocity at the perihelion of the corrected orbit by $v_{pc} - (v_{OE} + v_{CE}) = 2.765$ km s^{-1}.

We have performed the calculations for the epoch of 2641. Those calculations are, however, valid for any epoch. Our only concern is to choose the time of 1950 DA orbit correction such that at the perihelion of the corrected orbit the asteroid would approach the Earth. Such a problem was previously considered in Smulsky 2008, where a launch time of a space vehicle intended to pass near the Venus was calculated. The calculations by Eq. (18) – (24) were carried out on the assumption that the orbit planes of the asteroid and the Earth, and the Earth equator plane, are coincident. The calculation method of Smulsky 2008 allows the calculations to be performed at an arbitrary orientation of the planes. In the same publication it was shown that, following the determination of the nearest time suitable for correction, such moments in subsequent epochs can also be calculated. They follow at a certain period.

In the latter strategy for making the asteroid 1950 DA a near-Earth satellite, a total momentum $m_a \cdot \Delta V = m_a \cdot (0.911 + 2.765) \cdot 10^3 = 5.77 \cdot 10^{15}$ kg·m/s needs to be applied. This value is 4.8 times smaller than that in the former strategy and 53 times greater than the momentum required for making Apophis an Earth satellite. It seems more appropriate to start the creation of such Earth satellites with Apophis. In Corliss 1970, page 189, it is reported that an American astronaut Dandridge Cole and his co-author (Cole and Cox 1964) advanced a proposal to capture planetoids in between the Mars and Jupiter and bring them close to the Earth. Following this, mankind will be able to excavate rock from the interior of the planetoids and, in this way, produce in the cavities thus formed artificial conditions suitable for habitation. Note that another possible use of such satellites mentioned in Cole and Cox 1964 is the use of ores taken from them at the Earth.

Although the problem on making an asteroid an Earth satellite is a problem much easier to solve than the problem on planetoid capture, this former problem is nonetheless also a problem unprecedented in its difficulty. Yet, with this problem solved, our potential in preventing the serious asteroid danger will become many times enhanced. That is why, mankind getting down to tackling the problem, this will show that we have definitely passed from pure theoretical speculations in this field to practical activities on Earth protection of the asteroid hazard.

CONCLUSION

1. Through an analysis of literature sources, deficiencies of the previous calculation methods for asteroid motion were revealed.
2. The new method was used to numerically integrate non-simplified motion Equations of asteroid, the planets, the Moon, and the Sun over a 1000-year period.
3. On 21 hour 45' GMT, April 13, 2029 Apophis will pass close to the Earth, at a minimum distance of 6 Earth radii from Earth's center. This will be the closest pass of Apophis near the Earth in the forthcoming one thousand years.
4. Calculations on making Apophis an Earth bound satellite appropriate for solving various space exploration tasks were performed.
5. The asteroid 1950 DA will twice approach the Earth to a minimal distance of 2.25 million km, in 2641 and in 2962.
6. 6At any epoch, the asteroid 1950 DA can be made an Earth-bound satellite by increasing its aphelion velocity by ~ 1 km s^{-1} and by decreasing its perihelion velocity by ~ 2.5 km s^{-1}.

ACKNOWLEDGMENTS

The calculations were carried out on the supercomputer of the Siberian Supercomputer Centre of Siberian Branch RAS. Krotov O.I. took part in calculations of the Apophis motion on the system Horizons.

REFERENCES

Bowell, E. 2008. The Asteroid Orbital Elements Database, Lowell Observatory, ftp://ftp.lowell.edu/pub/elgb/.

Bykova, L.E., Galushina, T.Yu. 2008*a*. Evolution of the probable travel path of asteroid 99942 Apophis. In: Proceedings of Intern. Conf. Near-Earth Astronomy 2007, September 3-7, Terskol, Nal'chik, Russia, pp. 48 – 54.

Bykova, L.E., Galushina, T.Yu. 2008*b*. Trajectories in the region of possible travel paths of asteroid 99942 Apophis. In: Fundamental and Applied Problems in Modern Mechanics. Proceedings of the VI All-Russia Scientific Conference September 30 – October 2, 2008, Tomsk, pp. 419-420.

Cole, D.V. and Cox, D.W. 1964. Islands in Space. Chilton Books, Philadelphia.

Corliss, U. 1970. Mystery of the Universe. Mir, Moscow [in Russian].

Duboshin, G.N. 1976. Celestial Mechanics and Astrodynamics: A Handbook, Moscow, Nauka [in Russian].

Emel'yanov, V.A., Merkushev, Yu.K., Barabanov, S.I. 2008a. Cyclicity of Apophis observation sessions with space and ground-based telescopes. In: Proceedings of Intern. Conf. Near-Earth Astronomy 2007, September 3-7, Terskol, Nal'chik, Russia, pp. 38 -43.

Emel'yanov, V.A., Luk'yashchenko, V.I., Merkushev, Yu.K., Uspenskii, G.R. 2008b. Determination accuracy of asteroid Apophis' orbit parameters ensured by space telescopes. In: Proceedings of Intern. Conf. Near-Earth Astronomy 2007, September 3-7, Terskol, Nal'chik, Russia, pp. 59-64.

Ephemerides of the Jet Propulsion Laboratory. USA. 2008. http://ssd.jpl.nasa.gov/?ephemerides.

Everhart, E. 1974. Implicit single-sequence methods for integrating orbits. Celest. Mech. 10, pp. 35-55.

Garradd, G.J. 2004. MPE Circ., Y25.

Giorgini, J.D. and 13 colleagues. 2002. Asteroid 1950 DA Encounter with Earth in 2880: Physical Limits of Collision Probability Prediction. Science 296, No. 5565, pp.132 – 136.

Giorgini, J.D., Benner, L.A.M., Ostro, S.I., Nolan, H.C., Busch, M.W. 2008. Predicting the Earth encounters of (99942) Apophis. Icarus 193, 1-19.

Grebenikov, E.A., Smulsky, J.J. 2007. Numerical investigation of the Mars orbit evolution in the time interval of hundred million. A.A. Dorodnitsyn Computing Center, Moscow. (In Russian, http://www.ikz.ru/~smulski/Papers/EvMa100m4t2.pdf).

Ivashkin, V.V., Stikhno, K.A. 2008. An analysis of the problem on correcting asteroid Apophis' orbit. In: Proceedings of Intern. Conf. Near-Earth Astronomy 2007, September 3-7, Terskol, Nal'chik, Russia, pp. 44 – 48.

JPL Small-Body Database 2008. Jet Propulsion Laboratory. California Institute of Technology. 99942 Apophis (2004 MN4). http://ssd.jpl.nasa.gov/sbdb.cgi?sstr=Apophis;orb=1.

Mel'nikov, V.P., Smul'skii, I.I., Smul'skii, Ya.I., 2008. Compound modeling of Earth rotation and possible implications for interaction of continents. Russian Geology and Geophysics, 49, 851–858. (In English, http://www.ikz.ru/~smulski/Papers/RGG190.pdf)

Melnikov, V.P., Smulsky, J.J. 2009. Astronomical theory of ice ages: New approximations. Solutions and challenges. Novosibirsk: Academic Publishing House "GEO". The book in two languages. On the back side in Russian: Mel'nikov, V.P., Smul'skiy, I.I. 2009. Astronomicheskaya teoriya lednikovykh periodov: Novye priblizheniya. Reshennye i nereshennye problemy. Novosibirsk: Akademicheskoe izdatel'stvo "Geo".(Information see http://www.ikz.ru/~smulski/Papers/AsThAnE.pdf).

Rykhlova, L.V., Shustov, B.M., Pol, V.G., Sukhanov, K.G. 2007. Urgent problems in protecting the Earth against asteroids In: Proceedings of Intern. Conf. Near-Earth Astronomy 2007, September 3-7, Terskol, Nal'chik, Russia, pp. 25-33.

Smirnov E.A. 2008. Advanced numerical methods for integrating motion equations of asteroids approaching the Earth. In: Proceedings of Intern. Conf. Near-Earth Astronomy 2007, September 3-7, Terskol, Nal'chik, Russia, pp. 54-59.

Smulsky, J.J. 1999. Theory of the Interaction. Novosibirsk, Novosibirsk State University, Research Center, UIGGM SD RAS. (In Russian, http://www.ikz.ru/~smulski/TVfulA5_2.pdf). Smulsky, J.J. 2004. The Theory of Interaction. Ekaterinburg, Russia: Publishing house "Cultural

Information Bank". (In English, http://www.ikz.ru/~smulski/TVEnA5_2.pdf).

Smulsky, J.J. 2007 A mathematical model of the Solar system. In: Theoretical and Applied Problems in Nonlinear Analysis. Russian Academy of Sciences, A.A. Dorodnitsyn Computing Center. Moscow, pp. 119-139. (In Russian http://www.ikz.ru/~smulski/Papers/MatMdSS5.pdf).

Smulsky, J.J. 2008. Optimization of Passive Orbit with the Use of Gravity Maneuver. Cosmic Research, 46, No. 55, pp. 456–464. (In English, http://www.ikz.ru/~smulski/Papers/COSR456.PDF).

Sokolov, L.L., Bashakov, A.A., Pit'ev, N.P. 2008. On possible encounters of asteroid 99942 Apophis with the Earth. In: Proceedings of Intern. Conf. Near-Earth Astronomy 2007, September 3-7, Terskol, Nal'chik, Russia, pp. 33 – 38.

Standish, E.M. 1998. JPL Planetary and Lunar Ephemerides, DE405/LE405. Interoffice memorandum: JPL IOM 312. F – 98-048. August 26. (ftp://ssd.jpl.nasa.gov/pub/eph/export/DE405/).

Tucker, R., Tholen, D., Bernardi, F. 2004. MPS 109613.

INDEX

A

B

C

D

E

F

N

O

P

R

U

V

W

X

Y